Contact From Beyond

Chapter 1

Contact From Beyond

BILL Simone

Published by BILL SIMONE, 2025.

CONTACT FROM BEYOND

First edition. January 23, 2025.

ISBN: 979-8230633051

Written by BILL Simone.

Also by BILL Simone

My Paranormal Visits A 14 Year Journey
Contact From Beyond

Table of Contents

I want to thank my daughter Angela Simone for her cover painting. Angela was editor and photographer of my first book, My Paranormal Visits A 14 Year Journey.

Thanks to my wife Karen Simone and Daughter-N-Law Stephanie Simone for their support and encouragement then and now.

January 14, 2019 First Line Forms. A solid straight horizontal line with flashing lights moving along this line. The flashing has a waving motion from the left end to the right end of the line. The movement on this line reminds me of a voice recording being picked up on a scope.

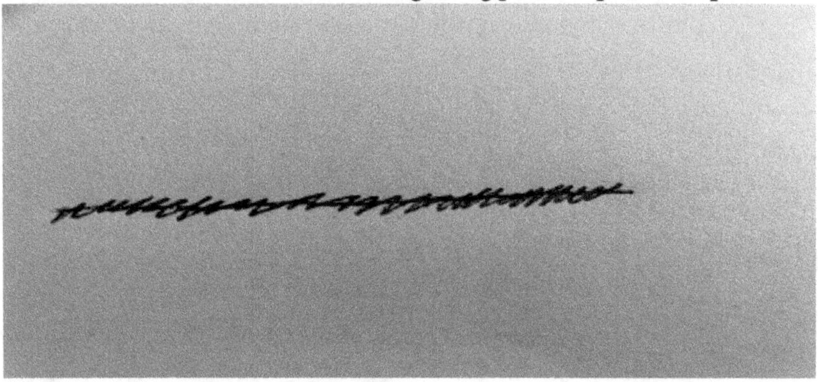

January 17, 2019 Rectangle Forms. The single line has now completely formed into the shape of a rectangle. The four sides are perfectly straight lines and also have the exact same flashing bright light on the rectangular shaped lines. The movement on the rectangle is a continuous flow from left to right going completely around the rectangular shape.

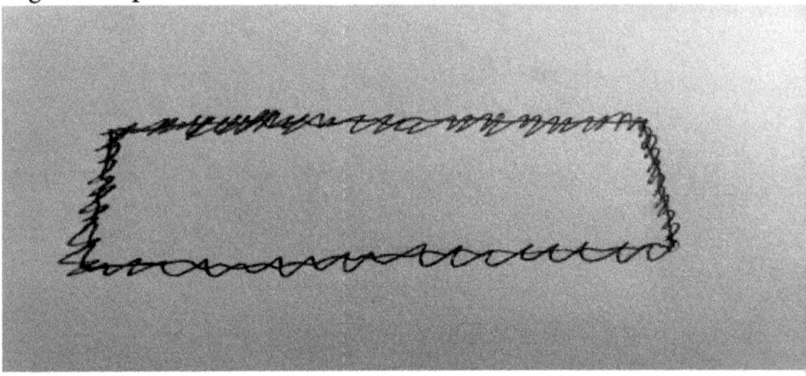

January 18, 2019 Three Square Boxes. I am watching the rectangle forming. I can see it is divided into three equal sections. It has become three perfectly shaped squares that are attached to each other. These solid lines also have flashing lights. But there is not any flowing movement, just the lights flashing. As I am watching this happen I see an image forming in each square but it's hard to see what it is, So I now say you have to come closer, I can't see what you are trying to show me. When I say this it moves closer and I can see a gravestone in each square. There is writing but I can't read it. The image then moves backwards and vanishes.

BILL SIMONE

Formed Tombstones Inside Each Square

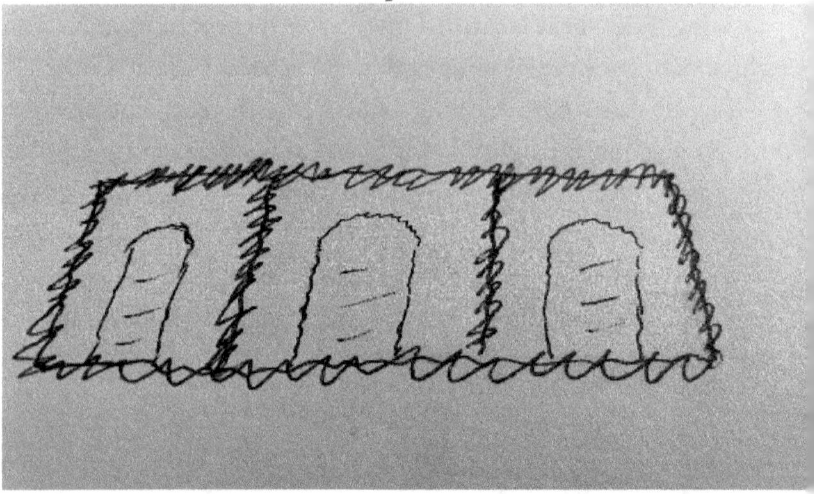

April 13, 2019 Line Appears. Suddenly a straight vertical line appears in front of me. There isn't any flashing or waveyness with this line. As I approach the line it starts moving back and slowly fades.

April 19, 2019 Thick Curved Line. A thick curved line instantly becomes visible, this line is very thick and dark. There is another line wrapped around the curved line.Both lines are stationary and are not flashing with any colors.

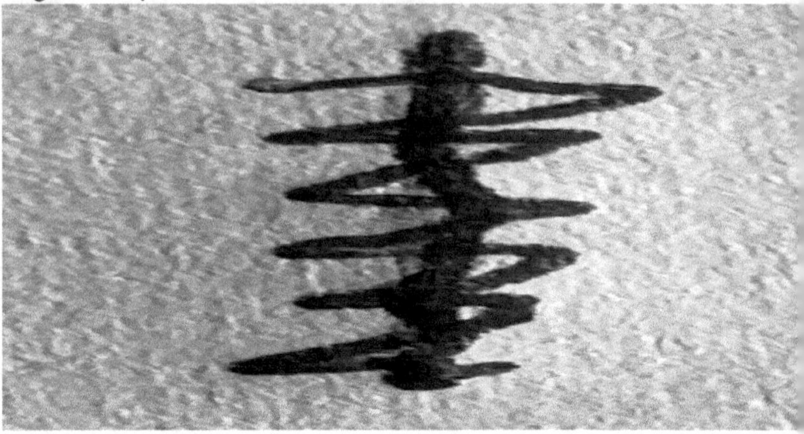

May 7, 2019 Bright Orb. I am opening my eyes and notice a small Orb moving quickly. This Orb is a bright white and is not an evenly round circle the size of a softball.

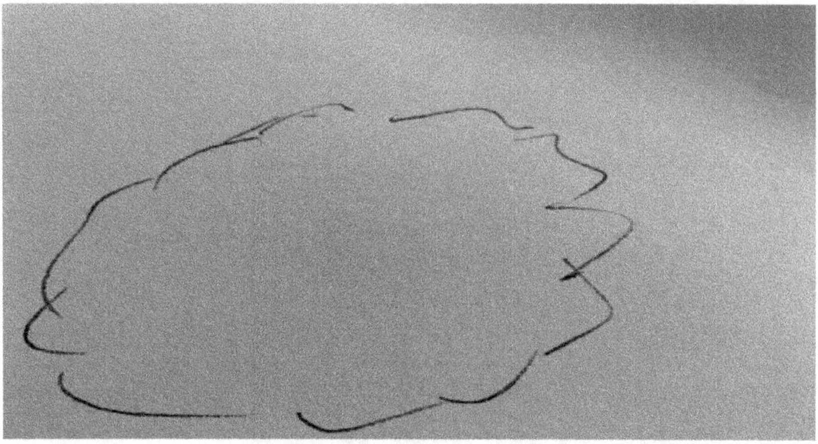

May 14, 2019 Two Bright White Orbs. As I looked up I noticed two bright Orbs going in opposite directions. Both of these Orbs are round and bright White. One Orb went through the left wall and the other Orb through the right wall.

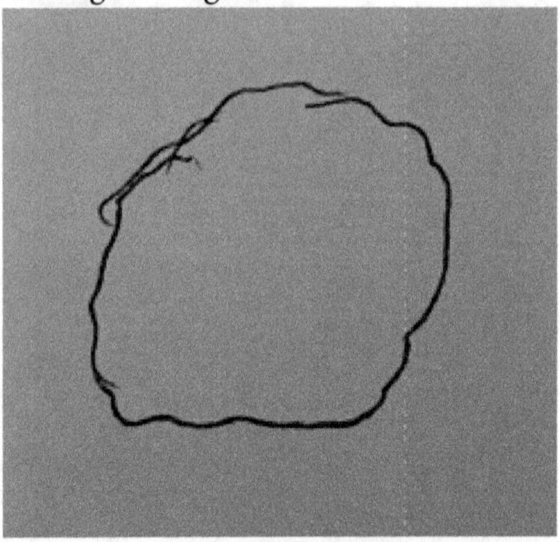

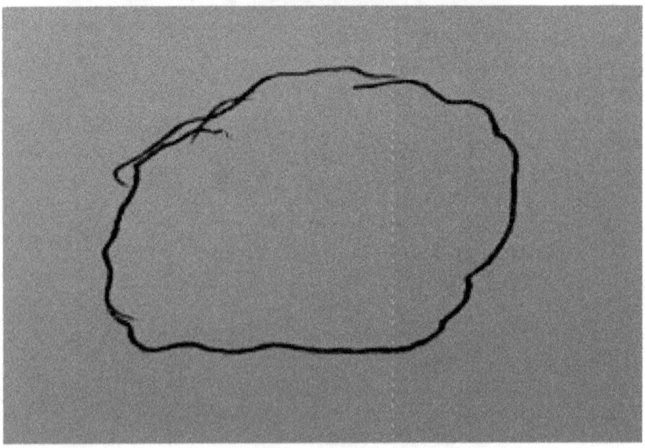

Thursday May 30, 2019 Floating Lines. Three vertical and waving lines equal in length are giving the appearance of vibrating. No flashing of any colors or of any other lines attached on them. The movement continues until they disappear.

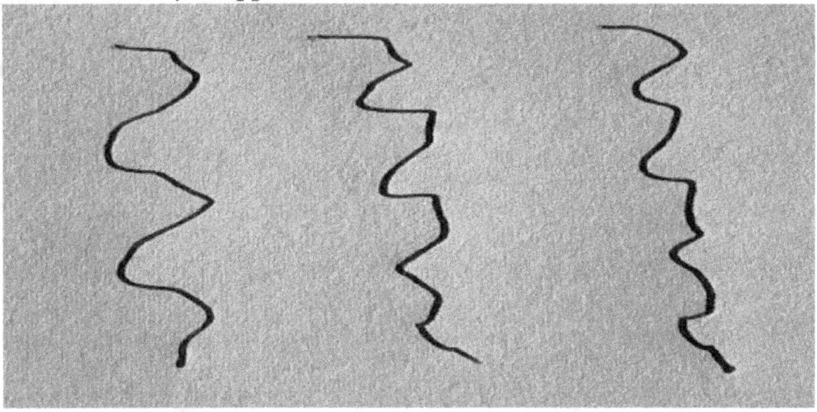

Sunday June 2 and 7th, Group Of Circles. The circles have caught my eye as the group moves very slowly across the room. They are attached to each other like a fence. Each circle has some type of movement around it. The same will happen June 7th.

Wednesday June 12, 2019 Two Vertical Rows Of Circles. I notice this group of circles floating upwards in front of the light where I am reading. The circles did not have any other movement around them. I could see through each circle.

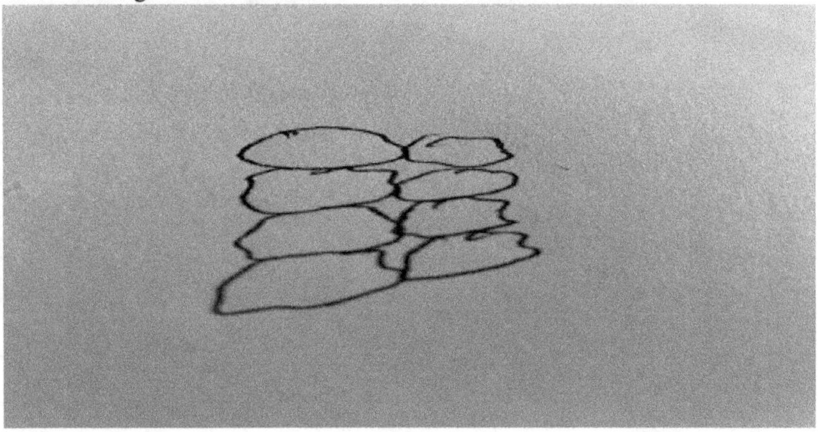

Thursday June 13, 2019 Four Straight Lines. I open my eyes and about six inches from my face are four straight vertical lines. No other movement or colors with these four lines that I was able to detect.

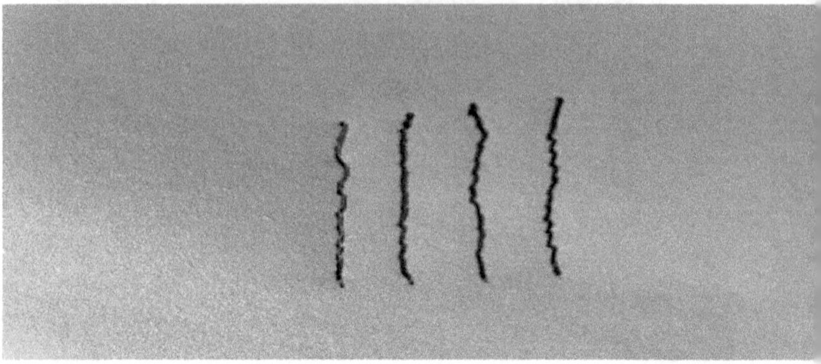

Friday June 21, 2019 Large Circle. The outline of this circle is very dark. It also has a flashing line going around it, but without any color. It stays stationary before moving away and slowly fading.

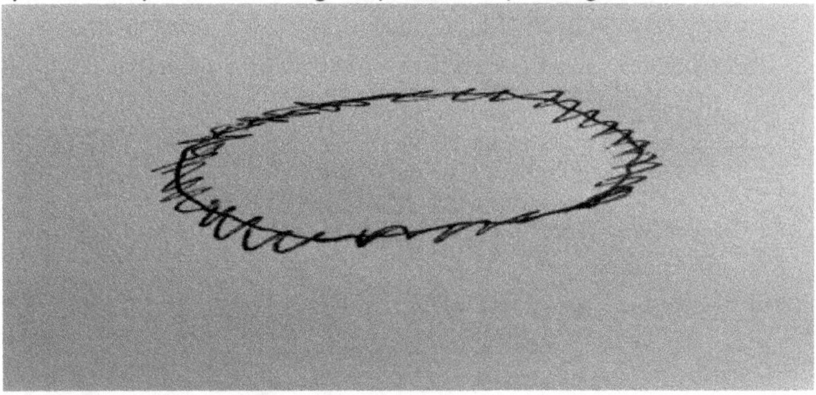

Sunday July 14, 2019 Two Objects. I noticed in the kitchen an Orb. The Orb is a bright Yellow-Orange color. This Orb is about two feet across. As I watch it move around the room it is changing shape. It is now a vertical image but has not lost any color. It is moving just like it did when it was in the Orb shape. Moves down the hall disappearing.

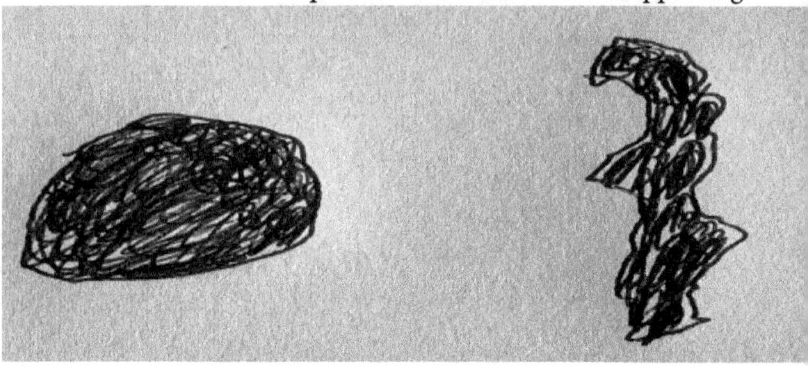

Friday July 19, 2019 A Group Of Circles. A small group of circles. The circles appear to be touching up against each other. There is a small circular movement around each circle. The movement is clockwise on each circle. The group is stationary, directly in front of me. It now moves backwards and slowly fades away.

Monday July 23, 2019 Circles Appear Again. The same group of circles I saw the previous day appear again. They are identical in shape touching each other's outer circumference. These circles also have the same clockwise movement of energy on them.

Friday July 26, 2019 Two Loaves Of Bread. As I wake up at three in the morning, why I don't know, I see two images. At first I thought of spiders hanging from the ceiling. They both are in front of me so I am trying to swing away at both of them. Now they are shaped into eyes. I swing again and they now have both elongated horizontally into the shape of bread loves. I ask what do you want and they both move backwards, but at different speeds and fade away.

Thursday August 8, 2019 Odd Figures. While reading I get a feeling of looking up. I don't hear anything, feel anything, just look up. Two stick objects clearly defined in shape without any movement and about one foot apart. No recognizable shape but are very close in their structures to each other.

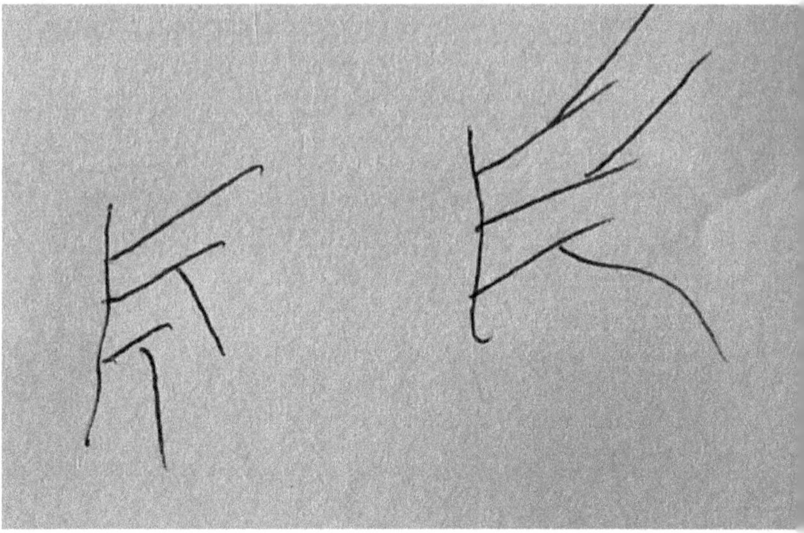

Saturday August 10, 2019 Bright Yellowish Orb. I see a Yellow-Orange Orb, it is about twenty feet in front of me.

Either it's dark in the center or it's a hole, but I can't tell. It has flashing broken dark lines around it but not touching the Orb. The Orb is moving around the room and is moving close to me. It now moves backwards and is smaller in size. I am talking to it and asking questions. It now moves into the kitchen smaller in size but still has a bright color. As I am watching for about fifteen minutes, I fall asleep not knowing what happened to the Orb.

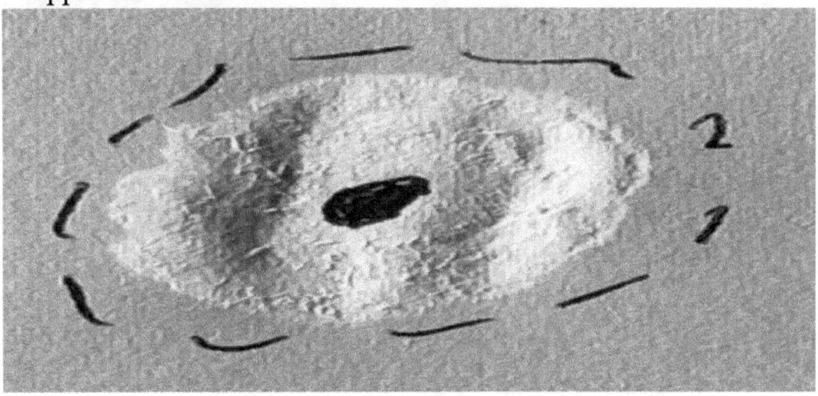

Friday August 16, 2019 Single Circle. When I open my eyes directly in front of me is a large single circle with some type of flashing movement going around the outline of the circle.

Saturday August 18, 2019 Horizontal Lines. Suddenly I opened my eyes and I saw about five curved horizontal lines. The lines are not moving or wavy in any way, but are curved in their shapes. I do not detect any flashing or other type of movement on these lines.

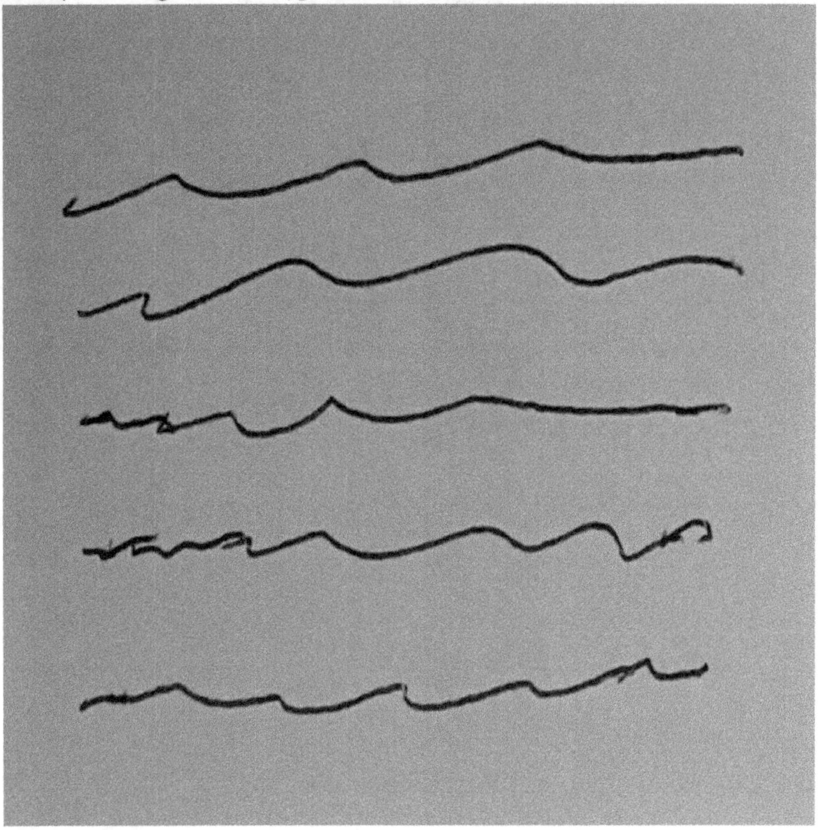

Sunday August 18, 2019 Vertical Lines. Second event today, these three vertical lines are moving across the room as I am watching TV. These lines are curved in shape and do not have any waving motion or flashing on them.

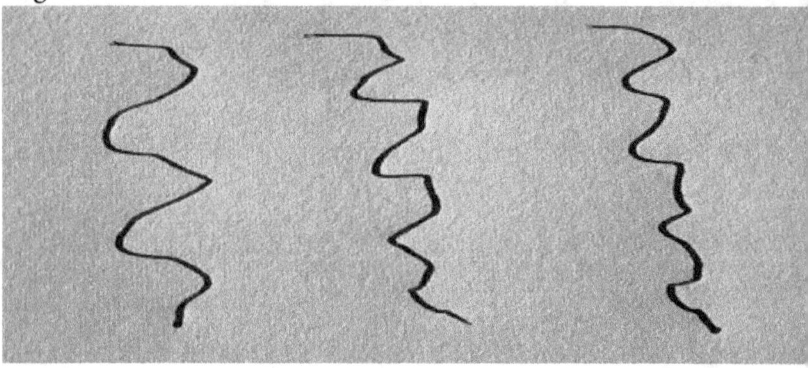

Wednesday August 21, 2019 Through The Ceiling. I see the same three vertical lines as they are moving across the room and upwards through the ceiling. Again these lines are not showing any waving motion, but are just curved in appearance. I did not detect any colors or flashing.

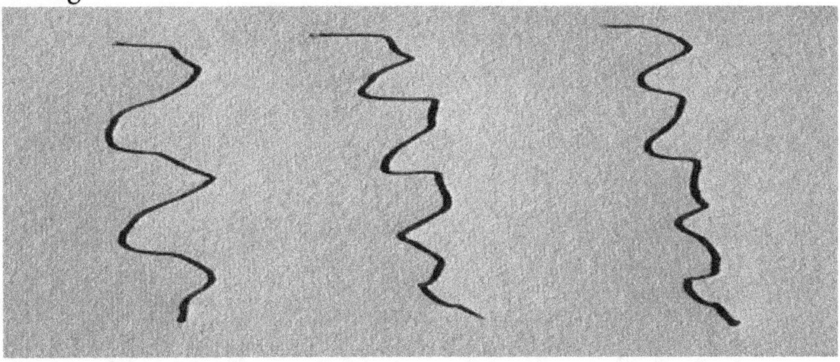

Saturday August 24, 2019 Eight Section Rectangle. I open my eyes from a deep sleep. I immediately noticed about five feet from the floor a rectangle perfectly sectioned into eight squares. Each square has a vivid dark outline. The squares outlines are not vibrating or have any flashing lights on them.

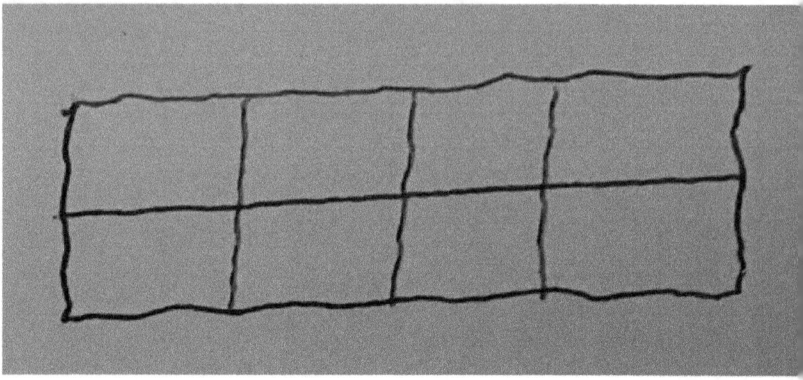

Monday August 26, 2019 Two Different Images As One. I am looking at a vertical image.The image has a design inside of it. On the outside are two solid vertical lines spaced about one foot apart. Inside these two lines a figure resembling the letter W or M on its side. The W or M figure has a vibrating movement with flashing light of White and Blue. I will see this visitor three times tonight within a five hour period.

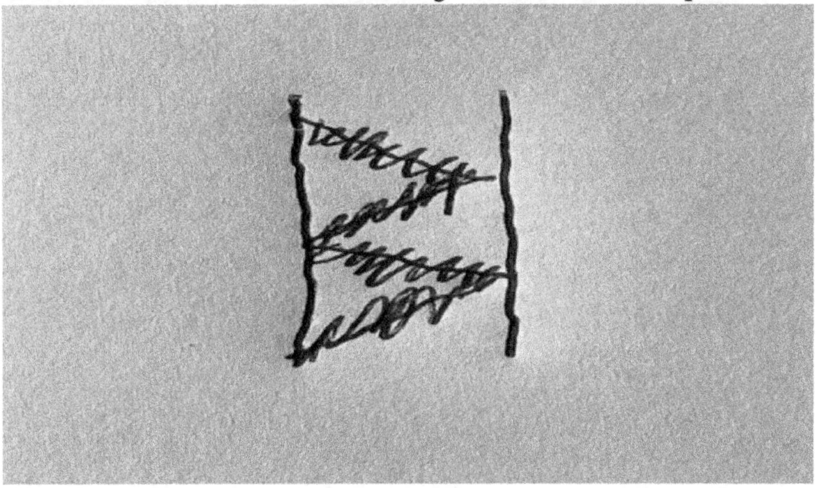

Tuesday August 27, 2019 Same Image. This is the exact way I had seen this visitor three times last night. One big difference with it tonight. The side lines were not flashing any colors or showing any vibrations. I will only see this image once tonight.

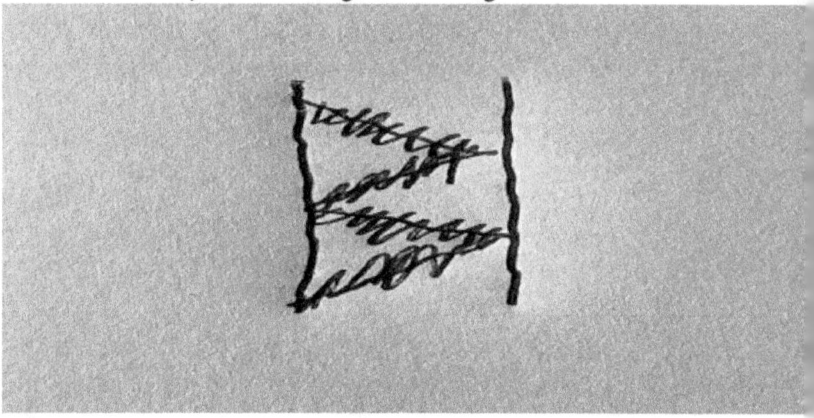

Saturday August 31, 2019 Six Vertical Lines. When opening my eyes I immediately notice six vertical lines. The two lines on the outside are about eighteen inches long and straight. Inside lines are a little shorter, about one foot in length. All four inside lines are wavy in shape and you can see them vibrating. None of the six lines had any colors or flashing of lights associated with them. This image will last about two minutes and then moves backwards slowly fading away.

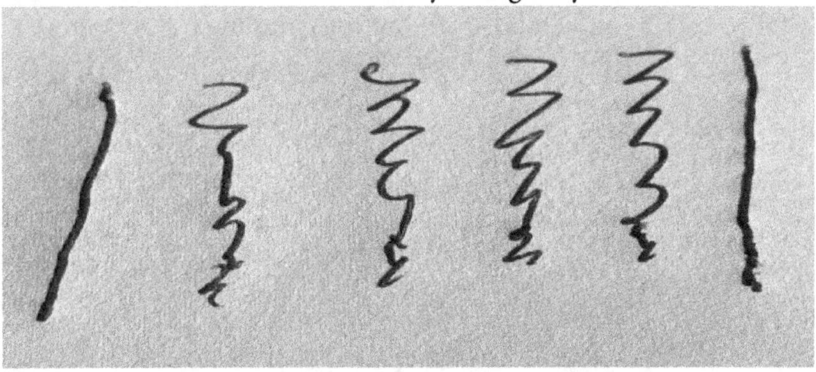

Wednesday September 4, 2019 Same Group Of Lines. I have seen this group of circles many times before. Between nine or twelve circles touching but not intertwined with each other. There is no vibration or colors associated with this group that I notice. It appeared that the group came out of the right wall in the bedroom. Their movement is quick and this is what catches my eye. As the group of circles cross the room they also have an upward climb and continue through the ceiling. I will see this group three times over the next five hours.

Friday September 27, 2019 Different Groups Of Lines.

One group of three curved vertical lines to my left. The three lines to my left side are pulasting with a vibration but don't show any colors on them.

The line to my right I have seen before with this pulsation along the main curved vertical line. It also was not showing any colors. Both vanish quickly with the one on left going first.

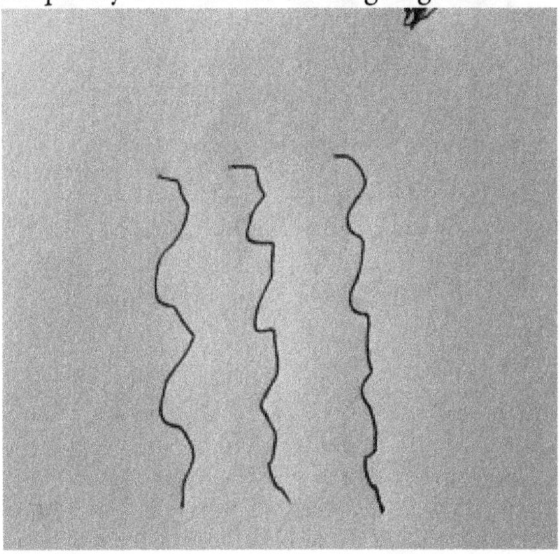

BILL SIMONE

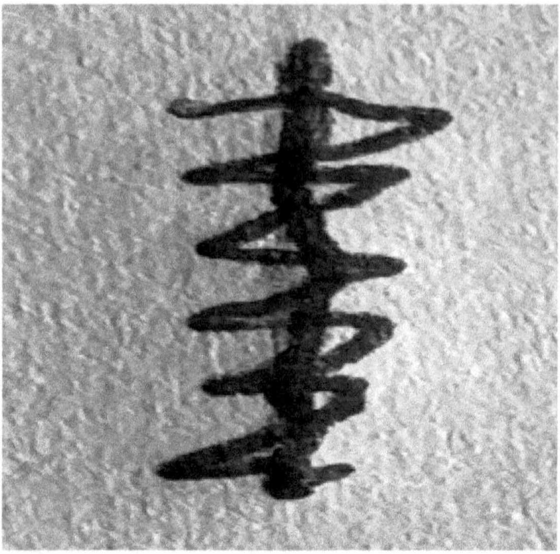

Wednesday October 2, 2019 Pulsating Vertical Lines. Four lines are a foot away from my face. I reach to touch the lines but my hand is just going through them. I can't feel anything, just air.

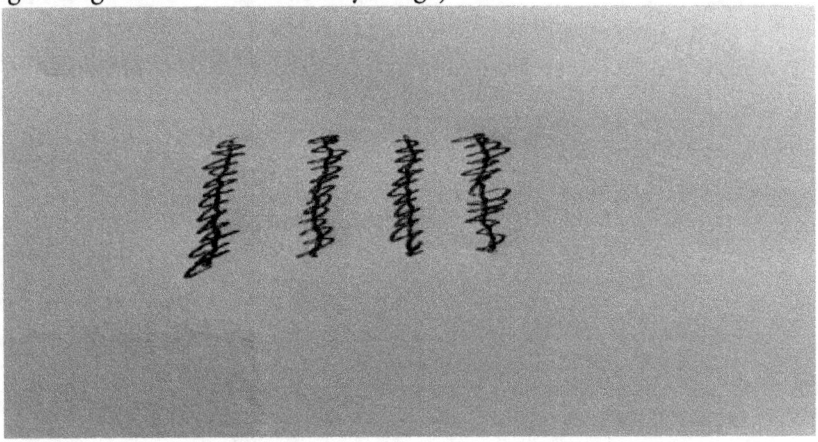

Friday October 12th, And October 16, 2019 White Mist. This occurrence will happen on two days but at different times. Eleven PM on the 12th and 1:00 AM on the 26th. Light is on while reading, during both days a mist is moving past the lamp causing a dimness in the room. As it leaves the dimness is gone.

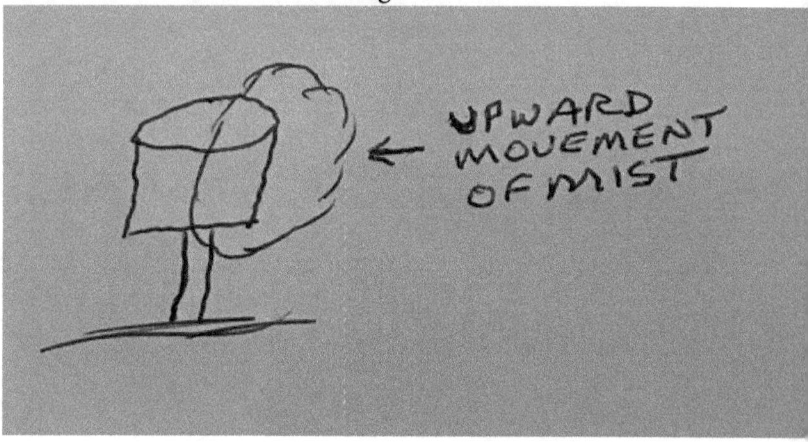

Wednesday October 23, 2019 Actual Photo. I feel a push on my shoulder. Just the nightlight is on and the pushing continues but I see nothing. I grab my cell phone and start rapidly taking nineteen pictures. Seventeen pictures are blank. One has a smear but no image. Then number nineteen in the upper corner there is an image. I cropped it and had it made into a 5x7 photo.

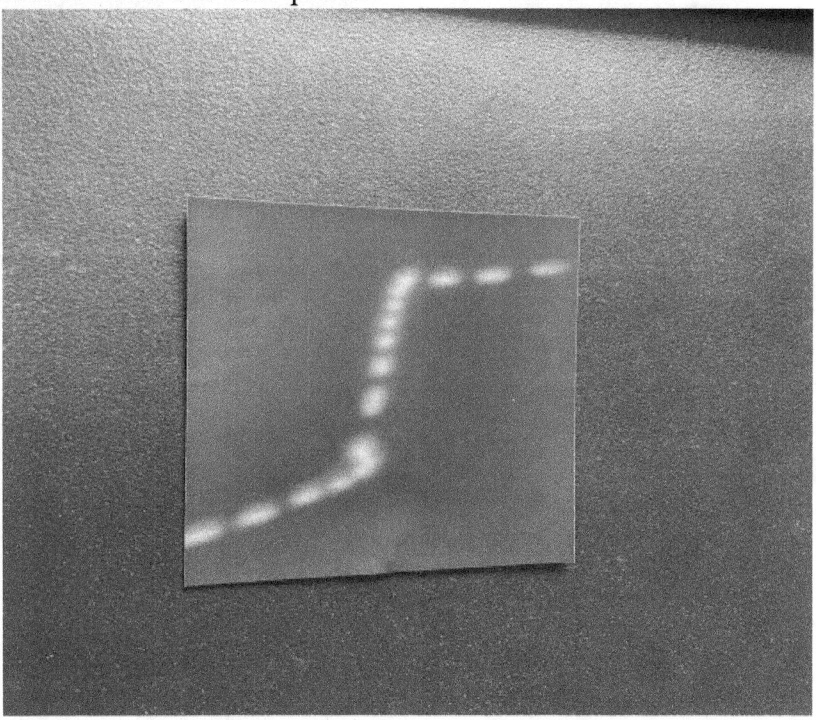

Monday October 28, 2019 Irregular Lines. About six inches from my face a large group of tangled lines. The lines are not making any recognizable form that I can see. The lines are pulsating as a group and not individually and without any lights or colors.

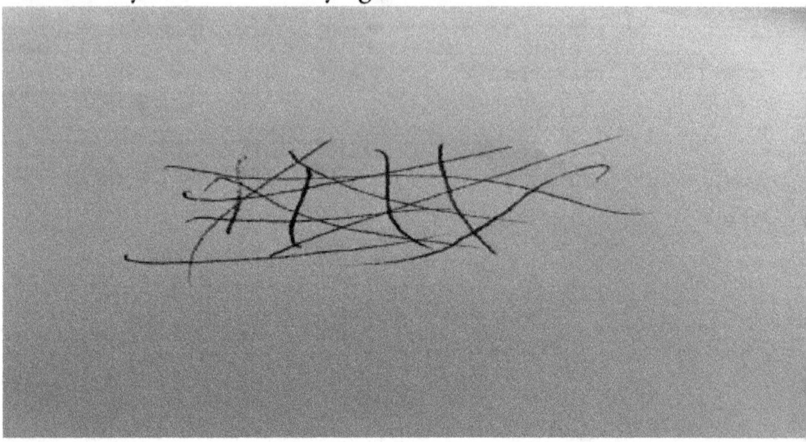

Thursday October 31, 2019 Irregular Flashing Lines. This same group of lines I saw on the 28th. But this time all the lines are not pulsating but have a bright White light glowing and flashing on them. This group stays about fifteen seconds and then is gone. Like a switch is turned off.

Monday November 4, 2019 Third Appearance. This is the third appearance of this group of lines. Each appearance is different in what they are doing. With this appearance the lines are pulsating and are flashing with a very bright White glow. With all three appearances the time frame is approximately fifteen to twenty seconds.

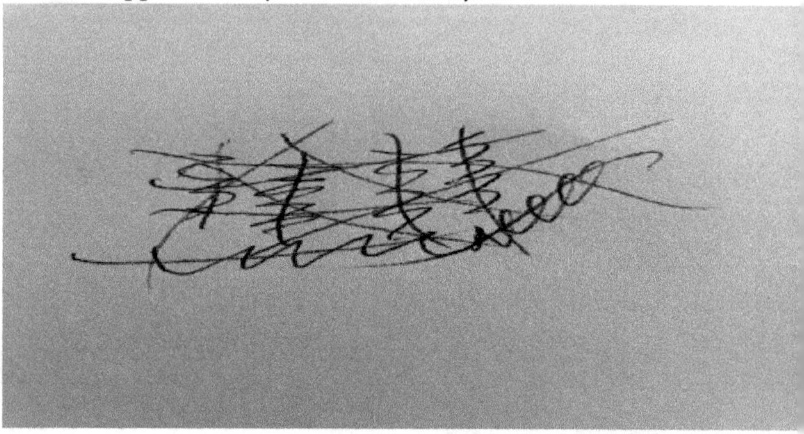

Tuesday November 5, 2019 Cross Figure. I see a single line moving slowly. The line is growing in size and is now taking a familiar form. It has now formed into a cross and has the Yellow Orange look. The color is flashing and I can see the dark line underneath the flashing color. It stays almost thirty seconds.

Tuesday November 12, 2019 Four Vertical Lines. Four very straight vertical lines are to my right. I turn my head to see them directly. All the lines have a dark curved line on them and are all pulsating at the same time. I do not see any lights or colors during this event.

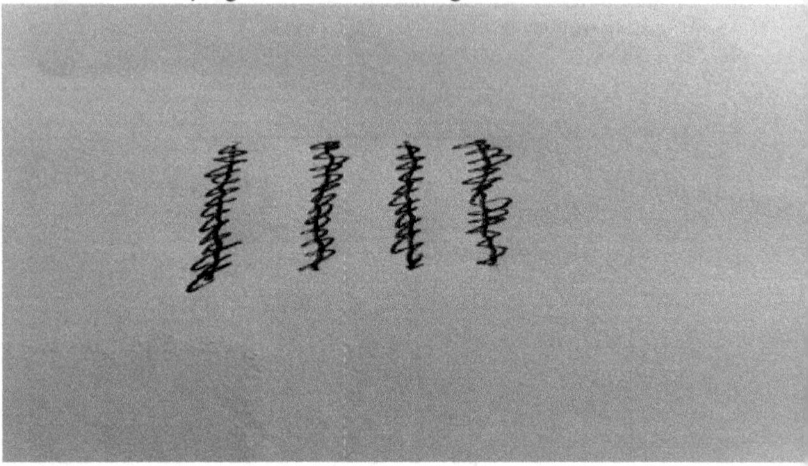

Wednesday November 20, 2019 And Sunday Nov.24th, Four Vertical Lines. These four vertical lines will appear exactly in the same manner as before. The days are different and the time of day is different. The first time on the 20th the lines are slightly to my right side. The second day they are directly in front of me. With both visits the lines had movement of flashing White lights moving rapidly up the lines.

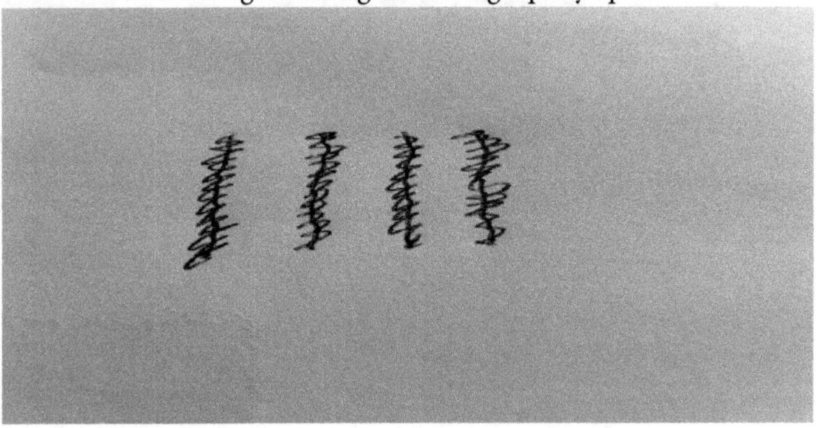

Saturday November 30, 2019 Two Orbs. From my room something gets my attention. Lifting my head to see what it is I see two very bright White Orbs flying around in the kitchen. Both are the size of a softball, flying around like moths around a light. They will do this for fifteen or twenty minutes before vanishing.

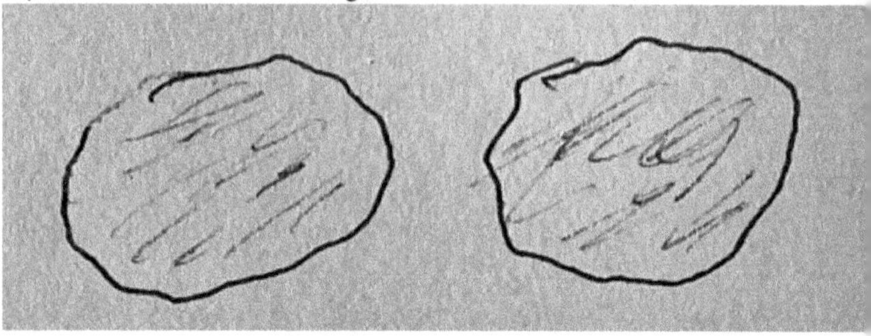

Wednesday December 10, 2019 Large Mist. I'm reading a book. The light in the room slightly dims enough to be noticed. As I turn to my left and look at the lamp, I notice a large misty area. It is moving upwards towards the ceiling. Without any pause in its movement it penetrates completely through the ceiling.

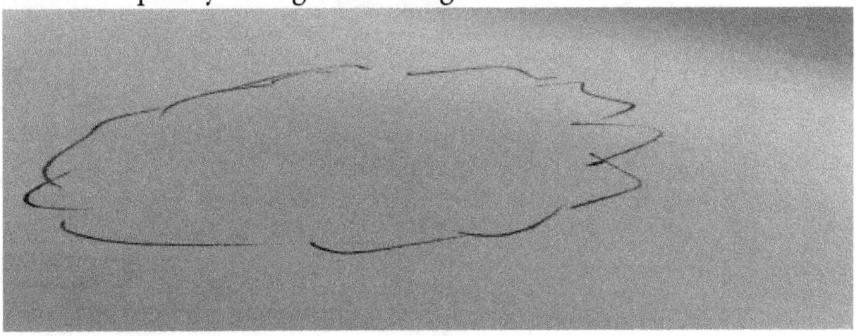

Saturday December 14, 2019 Lines Far Apart. I notice immediately how separated these lines are. They are far apart from each other. I then noticed a very thin line between the two outside lines. The outside lines start flashing a Greenish light, the movement is upwards through the ceiling.

Monday December 23, 2019 Pulsating Curved Lines. This object has touched me, which makes me look up from reading and notice straight ahead is a curved line, about a foot long. There is a pulsating line around it, but without colors or flashing of lights.

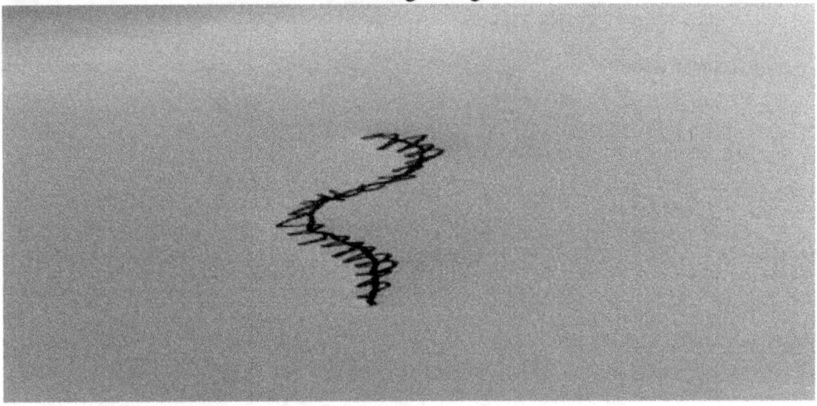

December 27, 2019 Twice Tonight. Twice tonight I will see this large group of vertical lines. This first time I noticed fast movement along the lines and had no color. Falling back to sleep and about fifteen minutes later they are there again. Why did I wake up? The size of the group is the same but the movement on the lines is slower. Again they are without any colors.

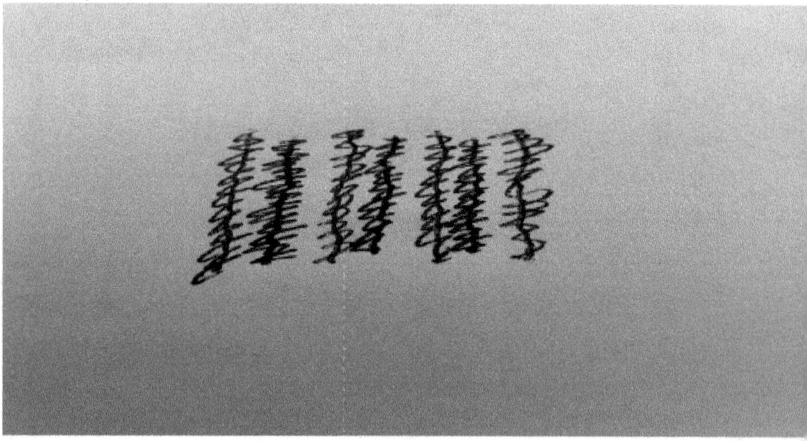

Chapter 2

January 15, 2020 Area Of Mist. As I am reading the light in the room dims. As I look over to the light I can see a shadow moving over the light. It's a large area of mist moving upwards. It is passing in front of the light causing the change in the room's brightness. The regular brightness returns as the mist passes.

January 16, 2020 Broken Lines. I notice a small area of vertical lines. The lines are not solid but broken up in their length. Each area of the line is flashing a White or Creamy color. The flashing is not in unison.

January 27, 2020 Rectangle With A Misty Center. I'm looking at this approximately 4x8 inch rectangle. The square has a milky White center and I can see cloudiness moving within the square. The square itself is stationary with a dark outline. As I get up I am a few inches away from it, I can put my hand underneath it. It now has the appearance of sitting on my hand, but my hand is just underneath it. As I walk around it there isn't any change, it looks the same. Nothing has changed with its structure or appearance. Then it just disappears as if it was turned off from a light switch.

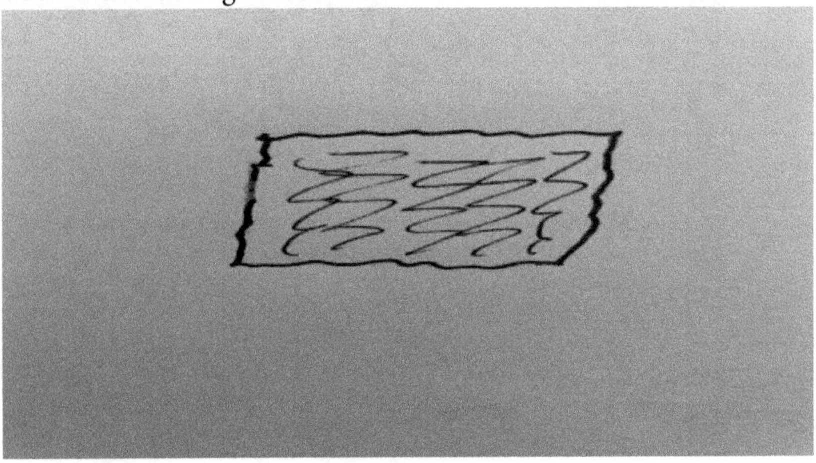

February 14, 2020 Lines And Circles. Two feet in front of me is a group of lines and circles. There isn't any pattern to them. They are tangled together and look like a tic tac toe board completely filled with x's and o's.

February 24, 2020 Stick Figure. This stick was moving from my left side to my right side. As it was passing it was only about three feet from my face. The lines had a motion but were staying in position with their connection. No flashing lights or pulsating that I was able to notice.

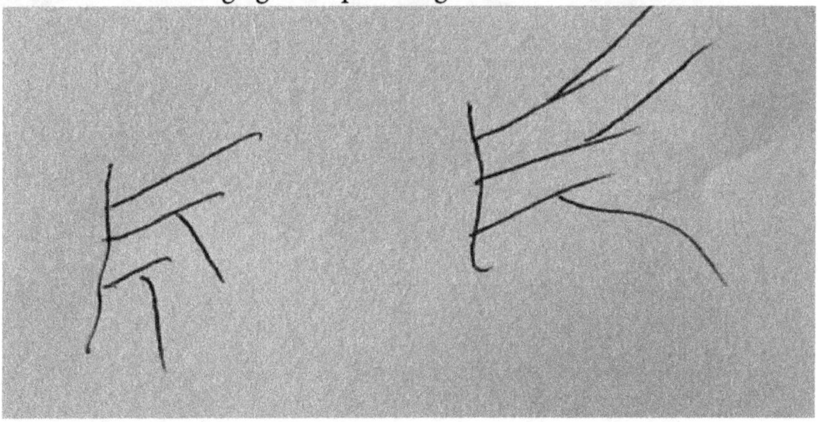

March 19, 2020 Chicken Wire Circles. This area is about two feet by two feet, about three feet in front of me. Small grouped circles that resemble chicken wire fencing. I'm able to see through each circle.

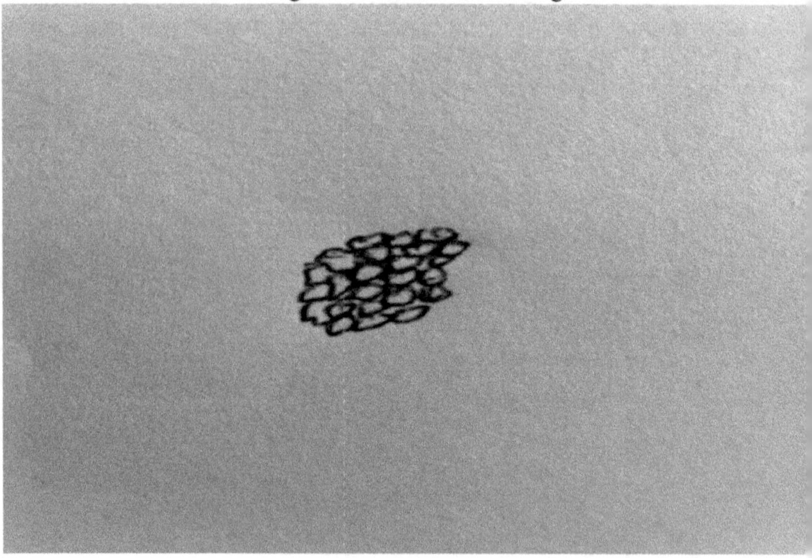

April 14, 2020 Mallet Image. Suddenly I noticed a very dark image about six inches in front of me. Its shape is changing into maybe a belt with a buckle at the end. I reach out for it and my fingers are going through it. Now staring at it, it's the shape of a mallet. It disappears as I turn the light on.

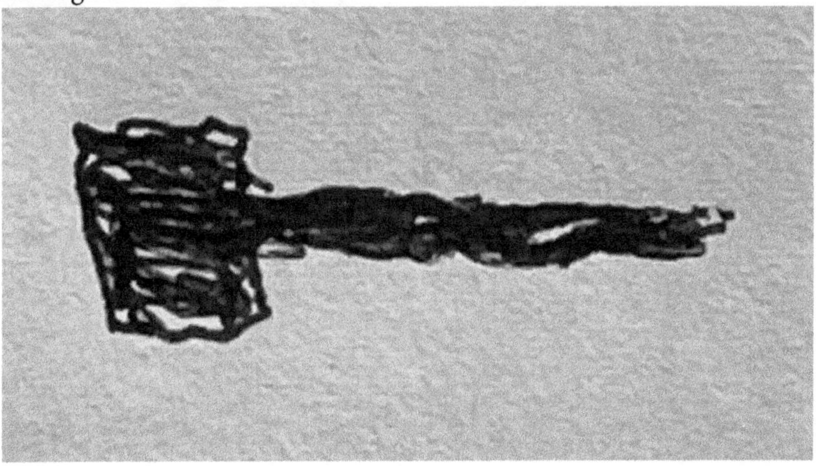

April 26, 2020 Intertwined Rectangles. Three times tonight, within a two hour period, this image will appear. It will be to my left in the first appearance and directly in front of me with the next two appearances. No flashing lights or pulsating of the lines that I noticed.

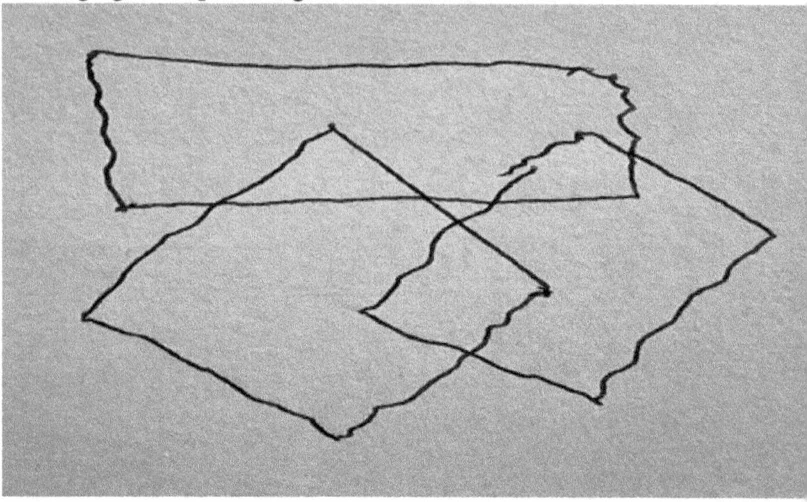

April 27, 2020 Flashing Image. What this image is trying to show me I do not know. It doesn't remind me of anything. It has come closer and is about two inches from my face. I say hi do you have a name? I do not get any response as I reach for it. My hand goes through it without feeling anything and suddenly it's gone.

April 29, 2020 Different Image. As with the April 27th event it suddenly appears a few inches in front of me. It has a slight pulsating movement, but without any flashing lights associated with it. I started talking to it, but never received any response.

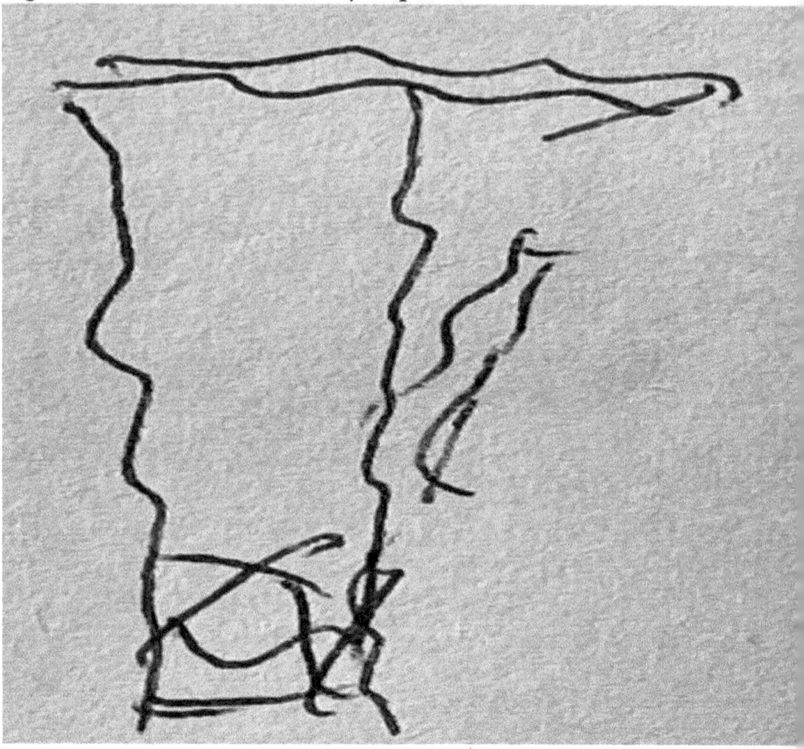

May 2, 2020 Four Elongated Circles. While reading, the table lamp to my left is on. A movement to my left catches my eye. I now see a circle moving upwards. It was even with the lamp shade when I noticed it. Now a second one is following, then a third and a fourth. All moving upwards about two feet apart from each other. Each circle has a movement resembling a ring of smoke rings from a cigarette.

May 11, 2020 Three Vertical Rectangles. I put my glasses on to look at this image. It is close and clear. I cover each eye individually to confirm the vision, it hasn't changed. I move my eyes from side to side, it does not move. I move my head in different directions, but it is still in front of me to see.

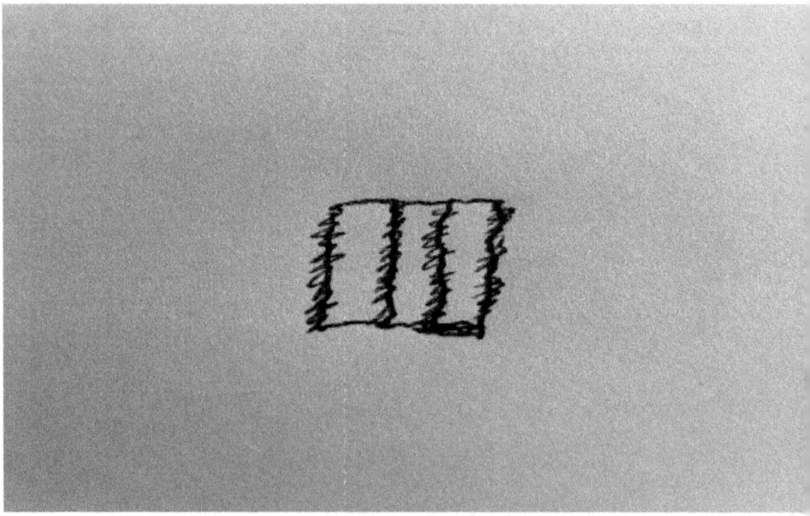

May 12, 2020 Two Objects. One of the objects is definitely an Orb. A White misty color inside the Orb.The other object is about five feet to the left of the Orb. A vertical object resembling a triangle with lines inside. Both of these images will appear twice tonight about three hours apart after the midnight hour.

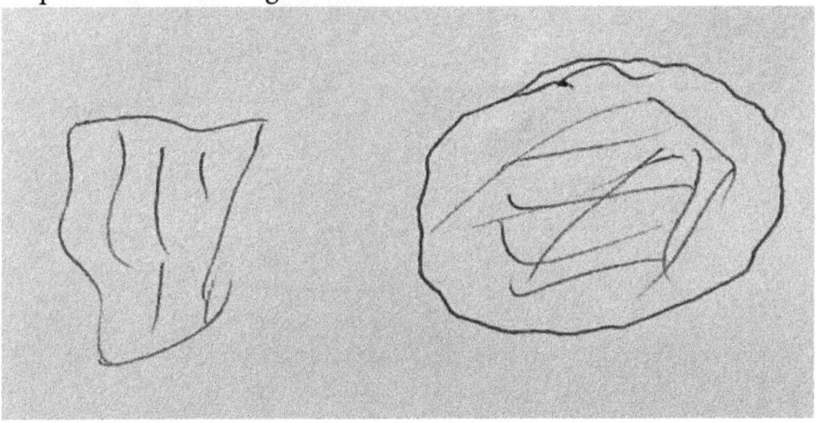

May 16, 2020 Two Horizontal Objects. Two bright streaks of light move past my face while using the computer. The top streak is about a foot in length with the lower one slightly shorter.

*May 17, 2020 A Forming Shape.*A cloudy area starts forming slightly to my right side. As I am staring I can see it is a rectangular shape with a dark outline and showing eight spots inside of the shape. Suddenly it moves backwards and fades away.

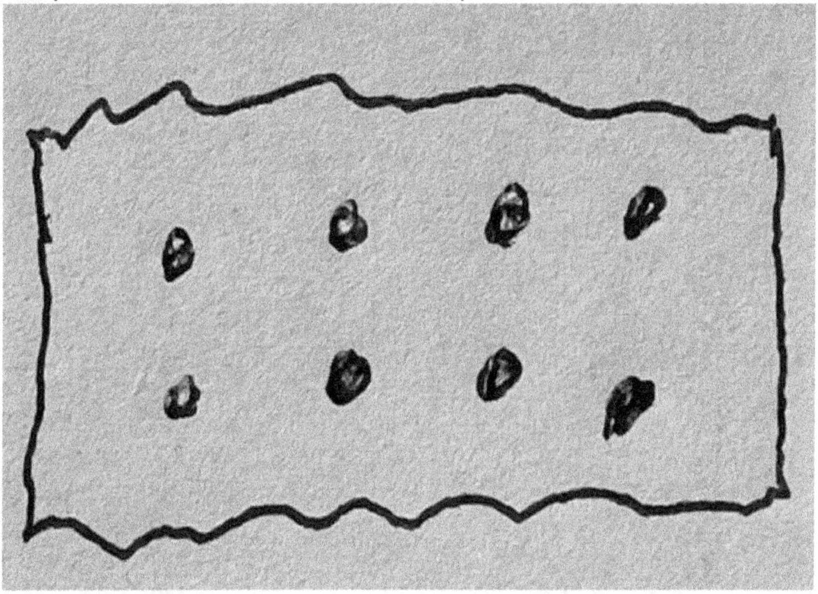

May 21, 2020 White Mist. Standing watching television, a mist to my left gets my attention. It's moving slowly across the room and towards the patio doors. As it now has reached and moves through the doors.

May 28, 2020 Changing Shapes. Watching TV with a nightlight on gives me some light in the room. To my left I see it. It's the same image I have seen before. How long it was there I don't know. It starts to move across the room in front of me. I can see it clearly change its structure three times while moving across the room before it slowly disappears.

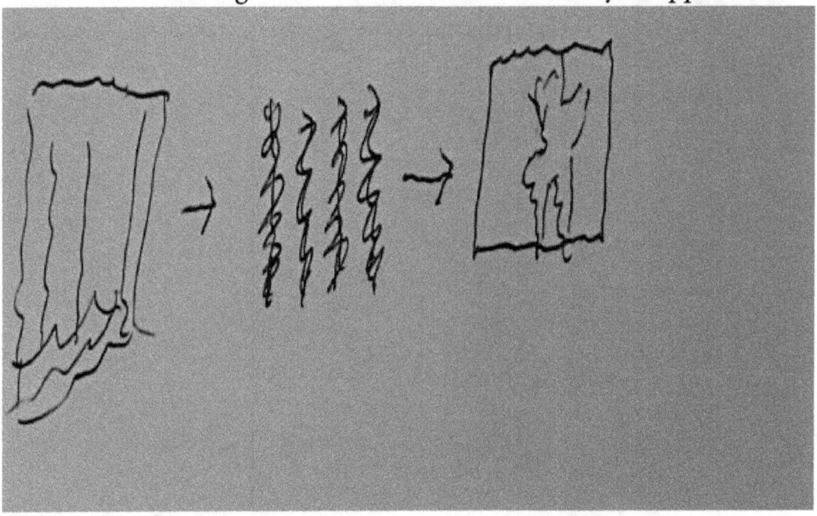

June 9, 2020 Unknown Energy. With a very odd shape I start talking to this object. I do not recognize the shape and it is very hard to describe. I start to cover and rub my eyes making sure my vision is ok. It is still only a few inches from my face. I Turn my head and do not see it. Return my head to a straight position and it's there in front of me. I say hi and reach to touch it. It moves back about half a foot and slowly fades.

June 12, 2020 Rectangle With Squares. When watching TV and the room is brightly lit, it pops up. Fully formed and easy to view in the room, a vertical rectangle with four squares inside of the rectangle. Just positioned in mid air, without any pulsation or flashing. All the lines with the rectangle and the squares are clearly defined. Then as quickly as it came, it quickly went as if a switch had turned it off.

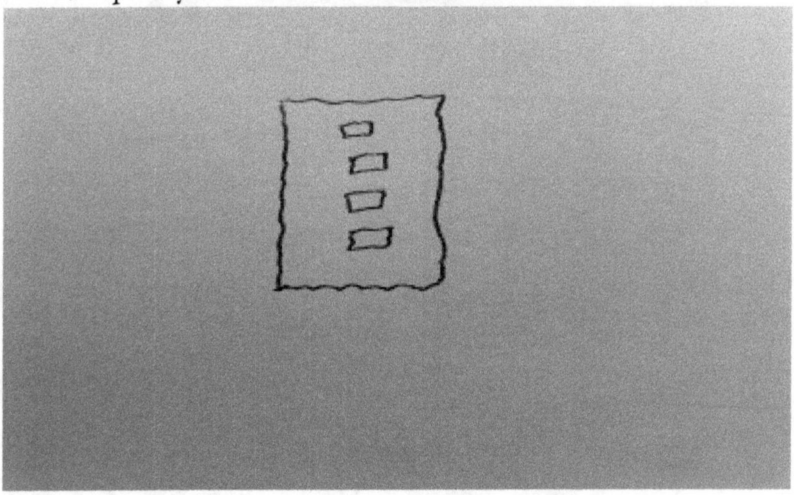

June 13,2020 Massive Group Of Circles. It reminds me of a group of frog eggs in the pond. The group is tightly bound together and is moving quickly to my right side and through the wall.

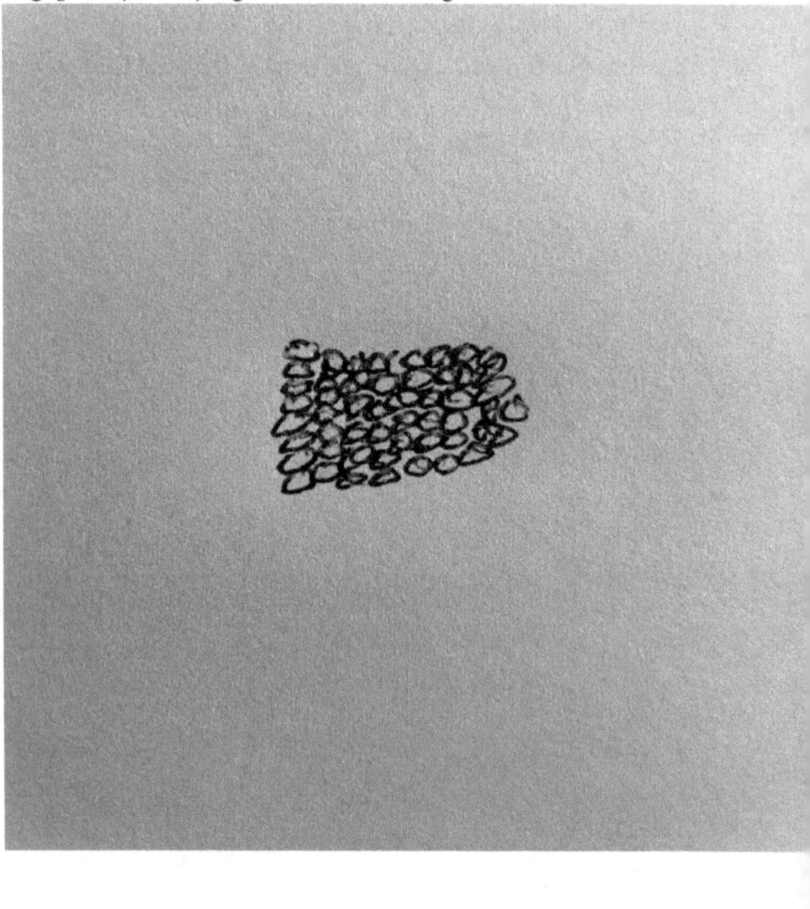

June 14, 2020 Kite. The kite image has returned. It is just about two inches from my face. I reach for it and the kite slowly moves back a few inches and disappears.

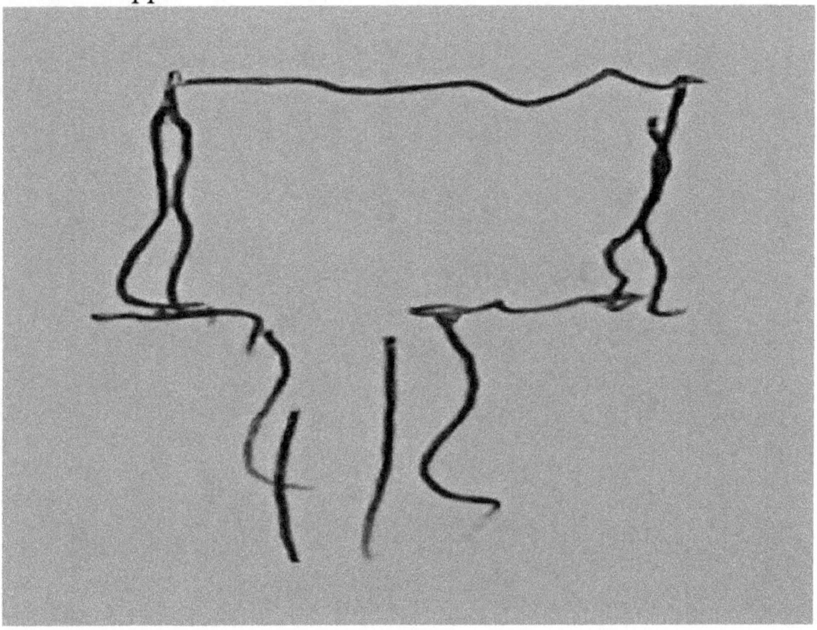

August 20, 2020 Two Streaks Of Circles. As I wake up I notice two separate lines of circles moving across the room in an upward and angled motion going through the ceiling.

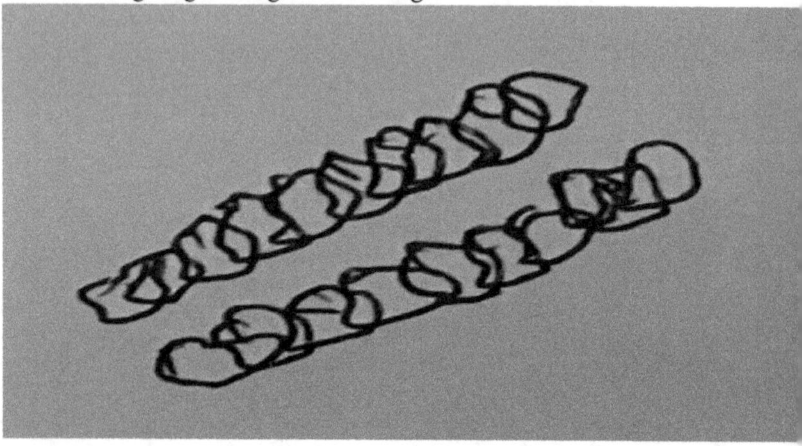

September 4th And 5th, 2020 Square With A Dark Center. Upon opening my eyes I see the outline of a square. The center is dark but not in any motion or flashing. I reach out and my hand goes through it without feeling anything. I turn my head, I can't see it. I return my head forward and it's still there. I reach again, and instantly it's gone.

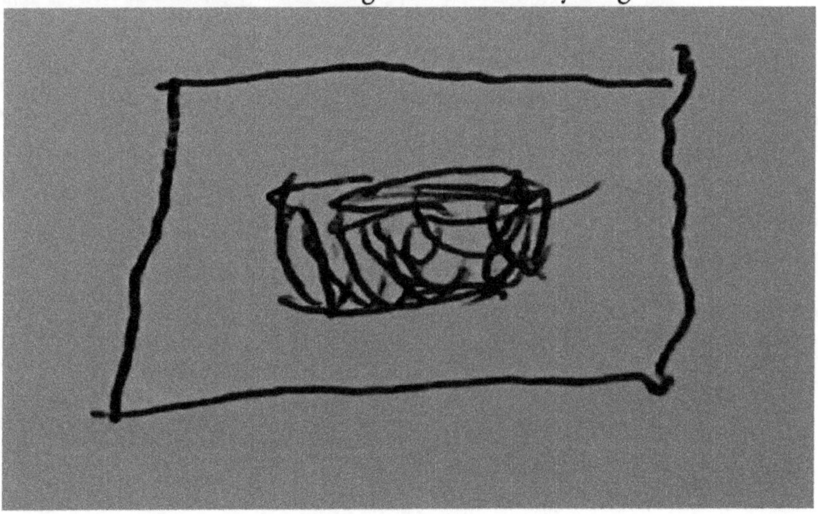

September 9, 2020 Mist. My friendly mist appears again moving upwards and past the lamp next to me and dimming the light. Upwards and through the ceiling without any hesitation.

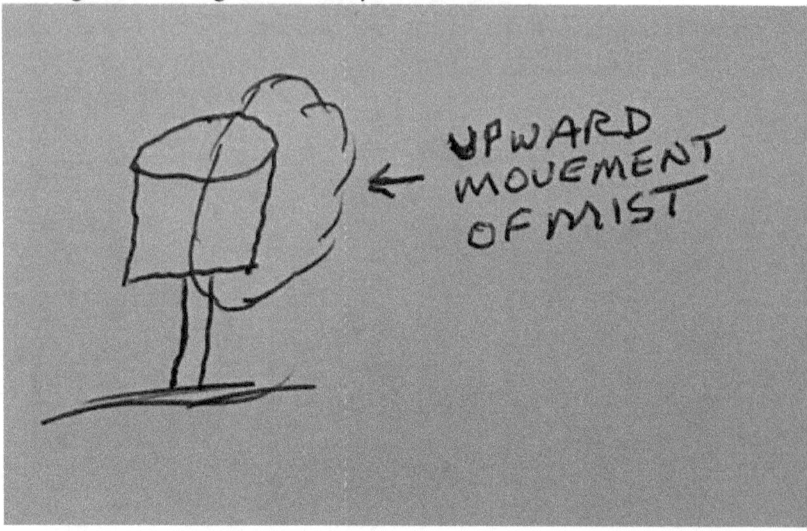

November 18, 2020 Mist With A Dark Center. The outline of this mist is moving in an uneven circular motion. I easily see a dark center, I have the table lamp next to me on. It is moving fast and making a right turn and moves up and through the ceiling.

December 14, 2020 Flashing Line. I see a horizontal line with movement on the line. No flashing colors, just a wavy movement. I rub my eyes to make sure it's not my vision seeing correctly. I finish rubbing and it's still there.

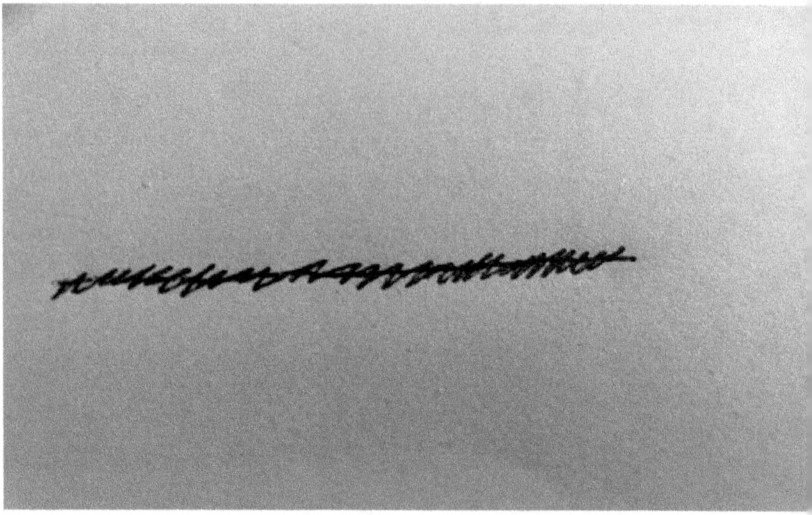

Chapter 3

January 4, 2021 A Face Appears. When opening my eyes I can see the outline of a man's face. I can't see through the Tan colored face. The eyes, nose, mouth and even the beard have a very dark outline.

January 20, 2021 Two Different Lines. I suddenly see two separate groups of lines. About five horizontal lines no flashing lights or pulsating, just straight lines.

Turning my head I see a small vertical line with another line wrapping around it. I now recognize this line, I've seen it many times in the past.

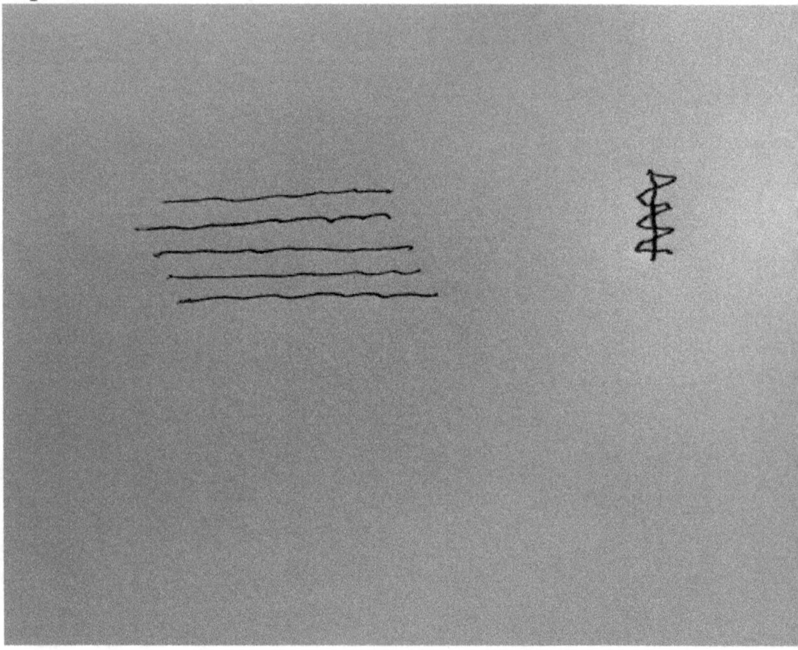

February 11, 2021 Familiar Face. I saw this face January 4, 2021. Seeing it again in identical appearance and shape. I will only see this face twice during the 14 year journey.

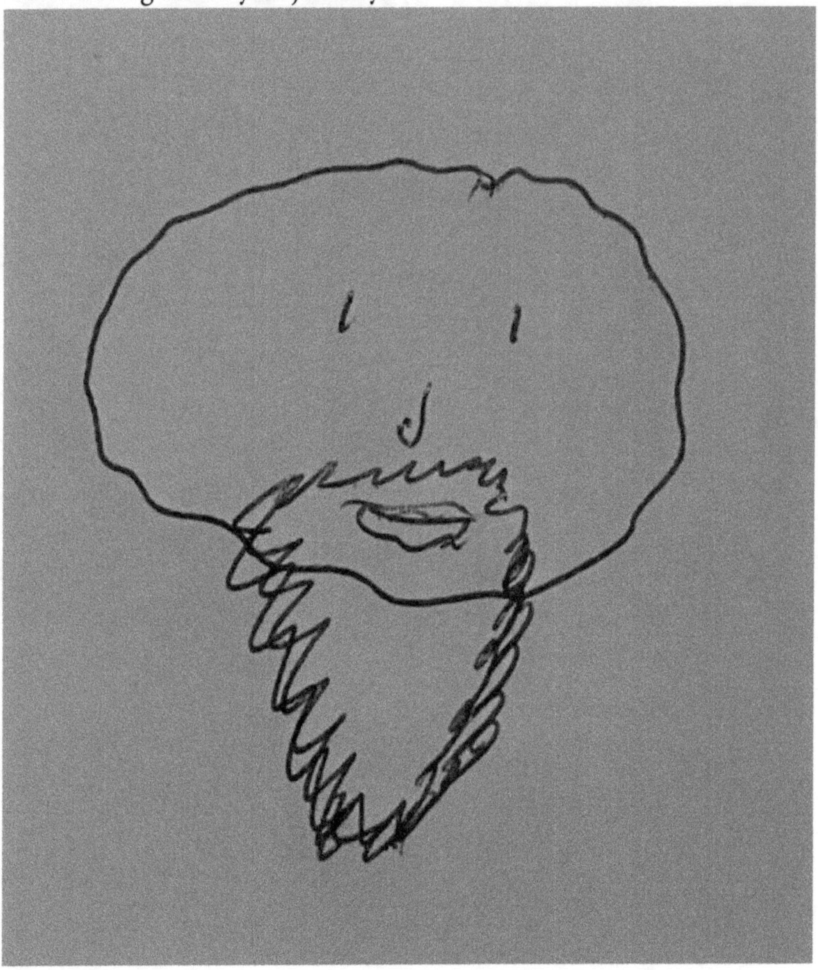

March 24, 2021 Solid Square. My kite returns with flashing and pulsating Blue and Green lights around its outline. The center is a solid black color and doesn't have any motion.

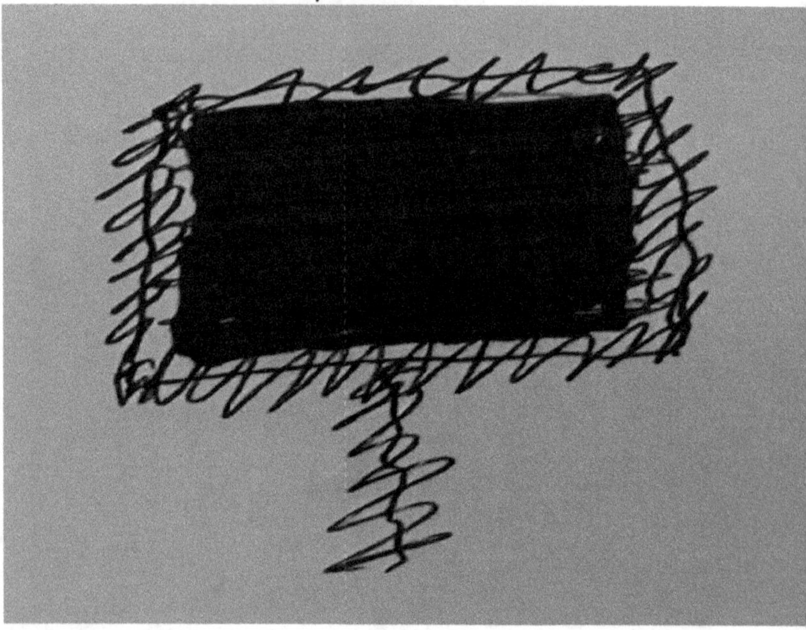

March 26, 2021 Rectangular Kite. Once again my kite returns. This time it's in a rectangle shape. It has the same flashing and pulsating lights moving around its outlines. This time the center has a slight mist with a few dark lines in the center.

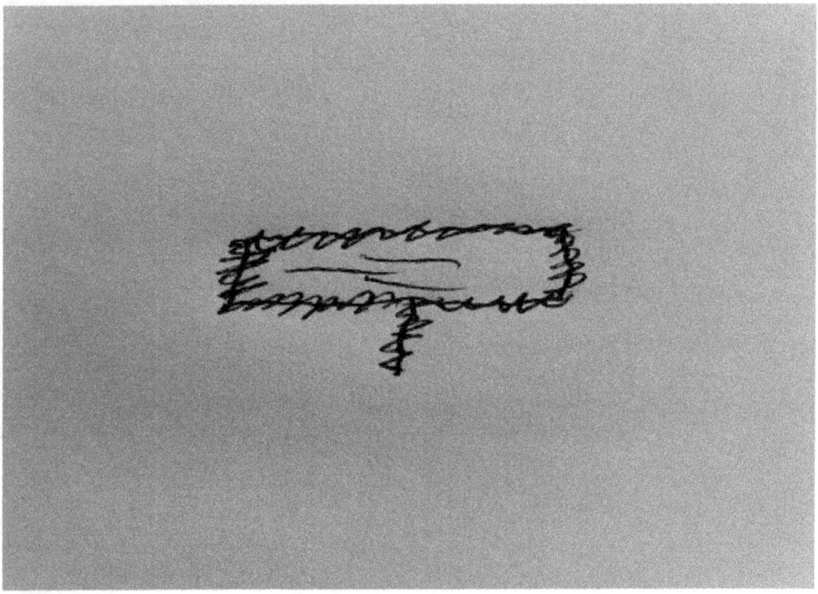

May 13, 2021 Solid Rod. Upon opening my eyes I notice a rod moving over my lap. It is not touching me as it moves across and is Brownish in color.

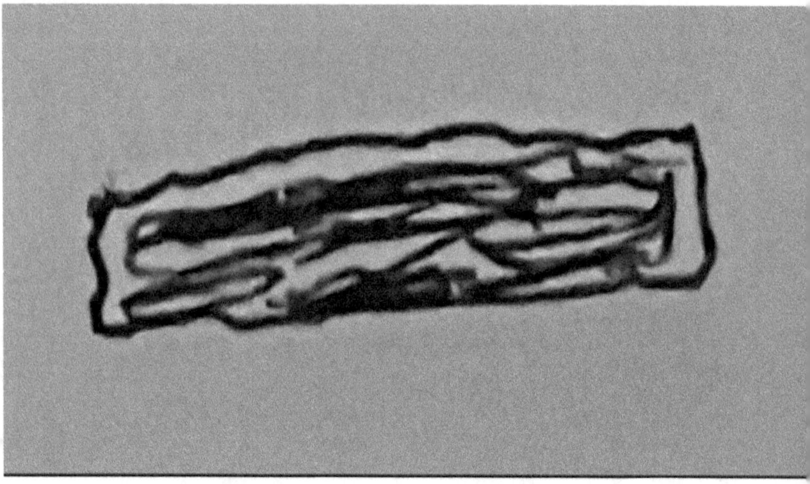

May 24, 2021 Circles And Mist. Two familiar objects will visit me tonight. My group of circles, but without any pulsating. About two hours later I will see a mist that will float by as it moves across the room and through the wall.

October 12, 2021 Friendly Circles. Each circle has a movement going around them. It looks like a flow of energy. This group just shows up in front of me and makes a sudden turn upwards and through the ceiling.

November 22, 2021 Curved Line. This line is not waving or flashing, but it is curved resembling a snake. It will keep its same shape as it moves across and disappears.

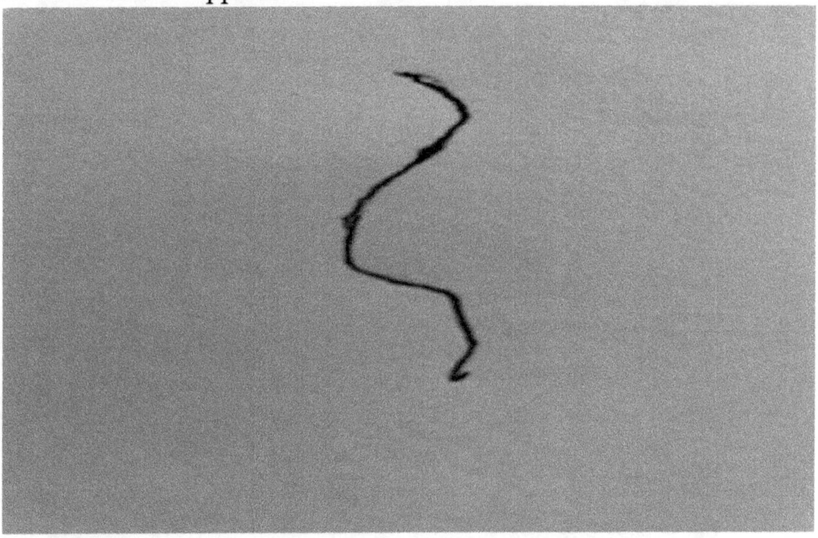

Chapter 4

February 23, 2022 Softball Orb. I have my eyes open and about two feet away from my face is a Green Orb about the size of a softball. You can see the Green color floating inside. It moves wildly in a circular motion and goes through the wall.

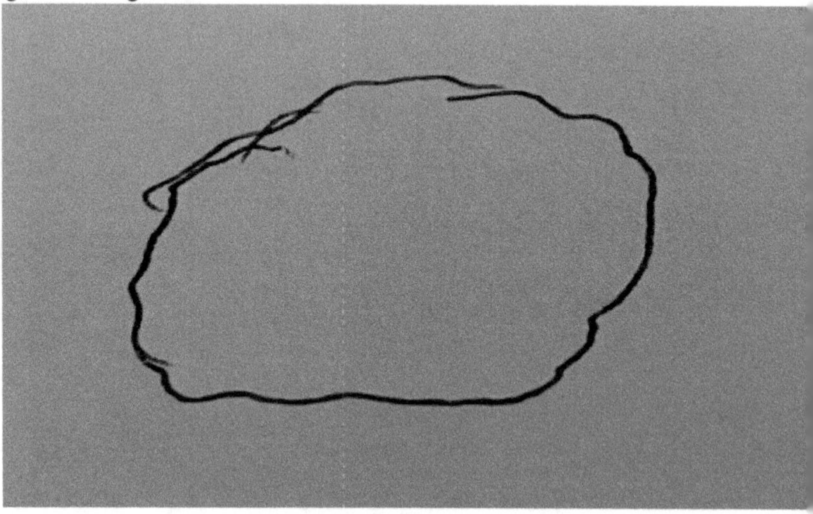

March 7, 2022 Lazy Eight. As I wake I see something floating up towards the ceiling, then it stops at the ceiling. It reminds me of the number eight laying flat. The outline has movement along it. It is pulsating like a signal on a meter. It stays in the same position for about ten seconds and rapidly moves through the ceiling.

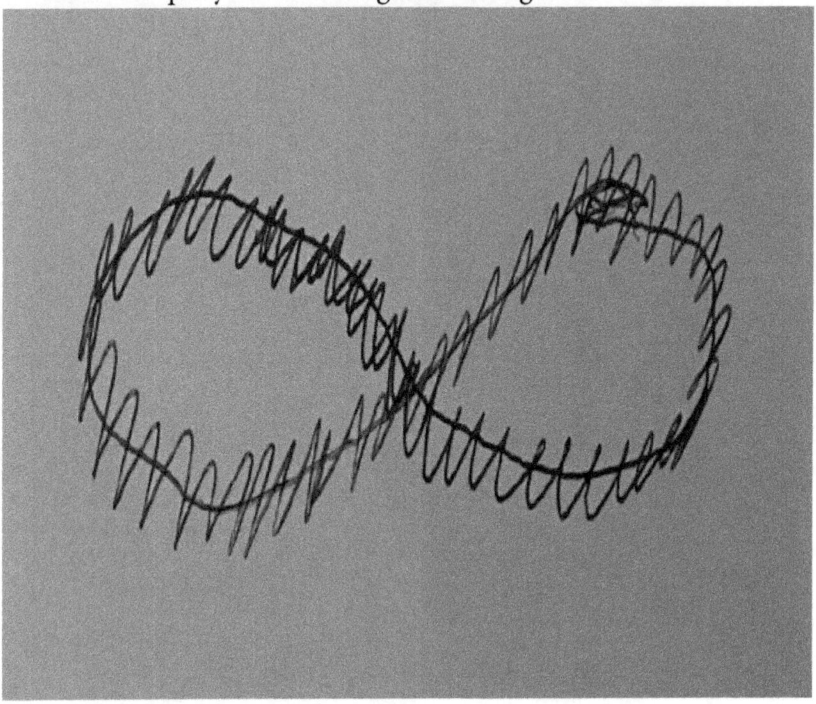

March 13, 2022 Flashing Lines. Going to the bathroom I notice four vertical flashing and pulsating lines. They seem to be about one foot in length and maybe ten feet in front of me. As I get closer they move back and slowly fade away.

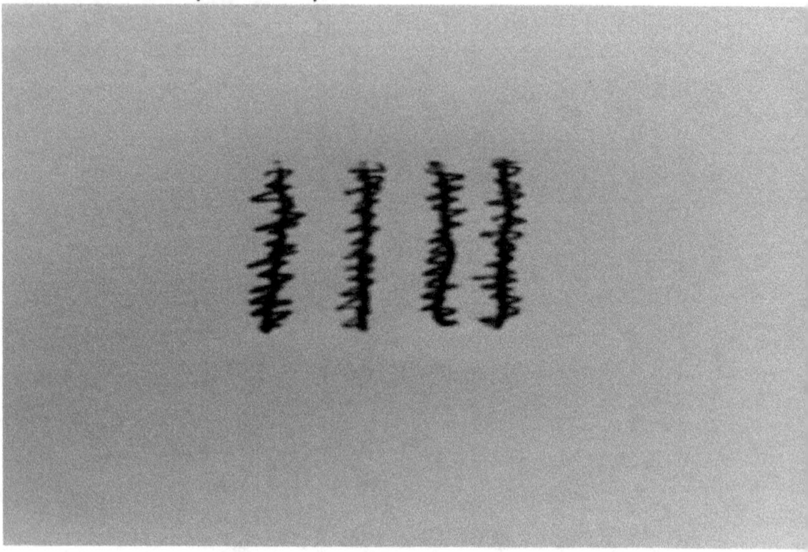

June 21, 2022 Tan Orb .I open my eyes from being awakened and there is my friendly Orb. Maybe two feet in front of me. Lightly tanned with the inside moving like a cloud. About ten seconds later it's gone, as if the switch is turned off.

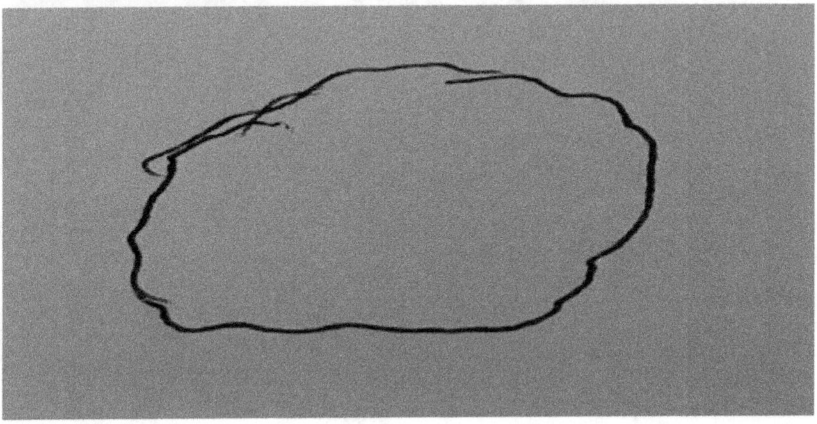

August 3, 2022 Grayish Tan Orb. My friendly Orb wakes me up again so I can see it and say hi. Unfortunately it will not stay long with its visit. Approximately fifteen seconds later it fades away.

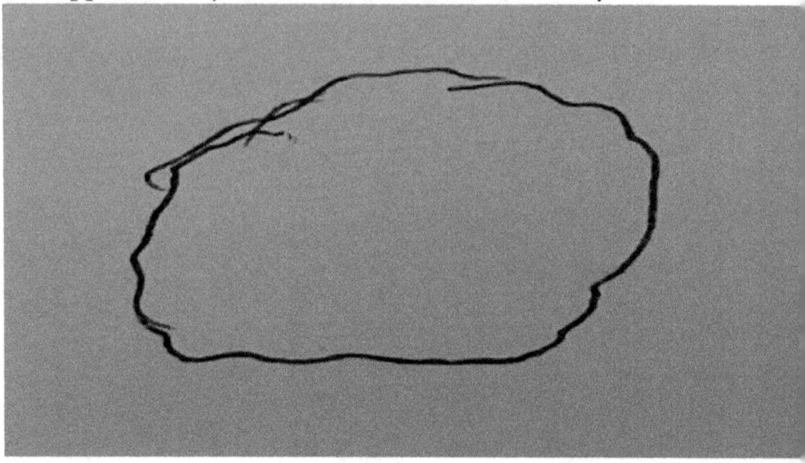

August 17, 2022 Irregular Lines. This event will happen twice tonight. The lines do not have any special shape. No wavy movement or flashing. They stay uniform as they move across the room and fade. Both events are identical in shape and their movements, but in different locations in the room.

September 10, 2022 Rectangle With Lines. Off to my left and up about five feet from the floor I see lines. It looks as if they are on a sheet of paper. It is moving closer to me now and looks as if there is writing on it. Not able to read what is written, I reach for it but can't grab it. My fingers are now going through the object and it disappears.

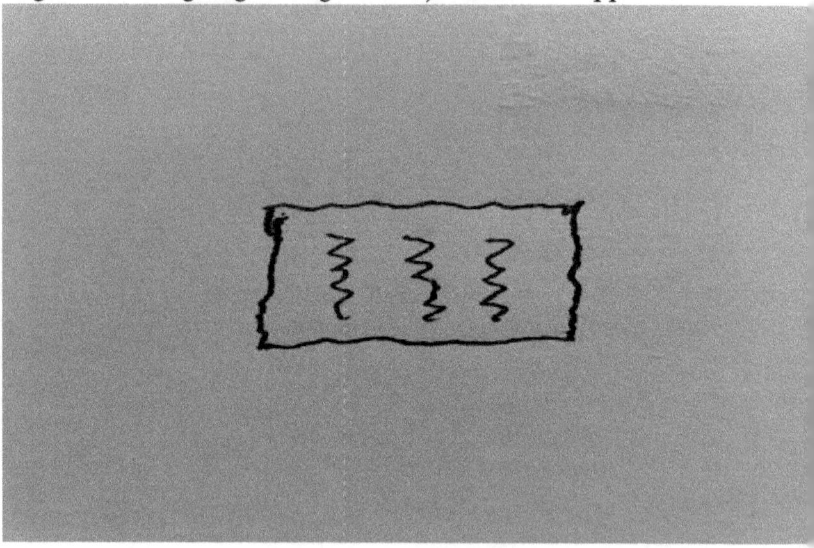

Chapter 5

February 4, 2023 Another Stick Figure. I have seen these lines before, but with different shapes and not as long. Straight lines with no flashing or lights. Just lines to the right side of a straight vertical line.

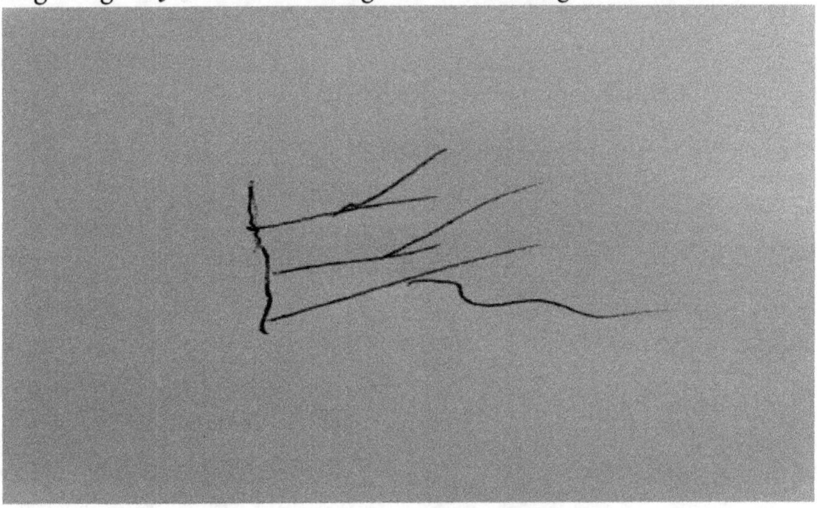

February 9, 2023 Mist Happens Again. With the light on it suddenly dims, slightly but enough for me to look over at the lamp. I notice the mist as it is slowly moving upwards and partially over the lamp shade. Upwards and through the ceiling.

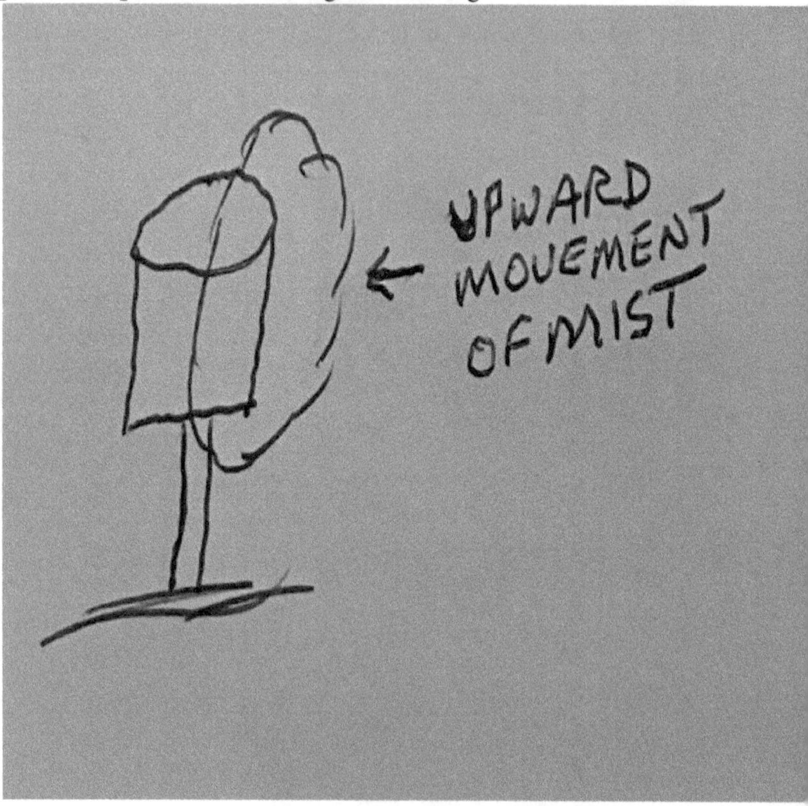

April 28, 2023 Curved Line. I am fully awake, I can see and feel this curved line pressing on my leg. As I slowly reach for it it moves backwards so I can't reach it. Becoming stationary and about five seconds later moves sideways disappearing.

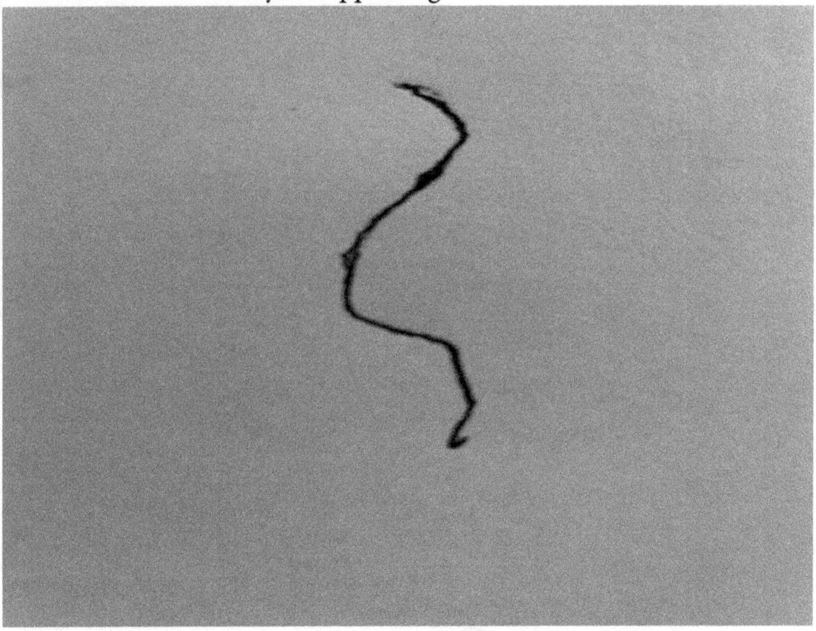

September 25, 2023 Black Square. There is a square directly in front of my face. Four solid square lines forming the square. Inside is a solid Black area but it is not filling the entire area. There is a clear area inside the square going around the inside of the four lines that form the box.

September 26, 2023 Rectangle With Squares. A rectangle is floating across the room. I lift my head up as it gets my attention. Two horizontal lines inside touching each side of the rectangle. I can see two short vertical lines intersecting the two inside horizontal lines. In the center is a dark circle. I am able to see through this image except for the dark circle area. It slowly moves to the other side of the room and through the wall.

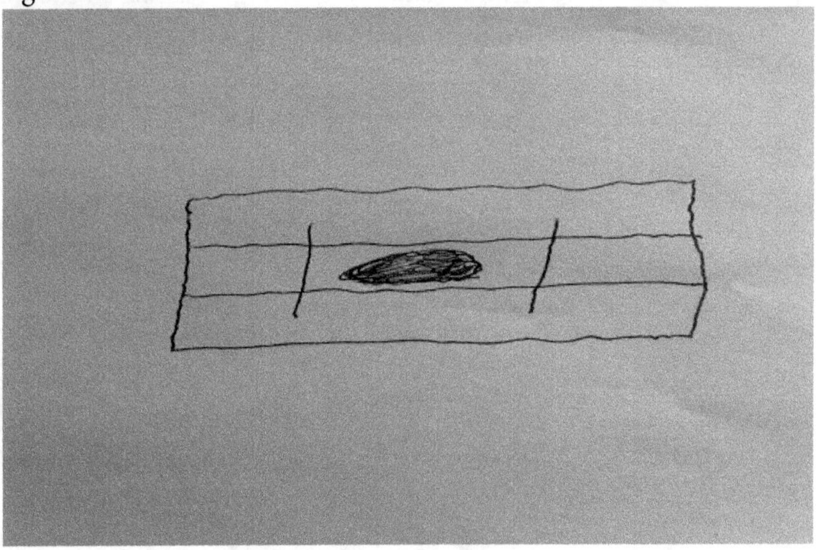

Chapter 6

April 1, 2024 My Large Friendly Mist. This is the last event that will be in the book for the year 2024.While I am reading it happens again. The light dims and when I turn to look at the lamp I see my usual mist friend. I get my K-11 meter and it is showing a reading all the way to the other end. The Red light is on. When the mist moves through the ceiling the reading immediately returns to its light Green color.

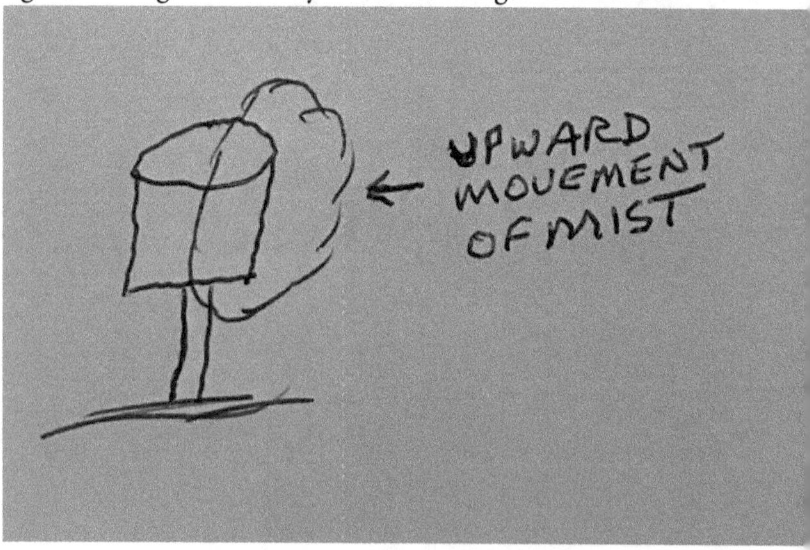

Summary

The images you have seen in this book are my own perceptions of what I was seeing.

Visit Paranormal Portal - Brent Thomas on UTube, Spotify, FaceBook, and wherever you get your Podcasts for several of my interviews, including Signs Of Spirits

M Paranormal Visits A 14 Year Journey

Contact From Beyond

The book is available in Digital or Paperback at many locations.

You can contact me via Email: baskfour04@gmail.com

Facebook and X

Bill Simone

Also by BILL Simone

My Paranormal Visits A 14 Year Journey
Contact From Beyond

www.ingramcontent.com/pod-product-compliance
Lightning Source LLC
LaVergne TN
LVHW012321130926
847902LV00027B/367